REVISE AQA GCSE
Mathematics A

REVISION GUIDE

Foundation

Series Consultant: Harry Smith

Author: Harry Smith

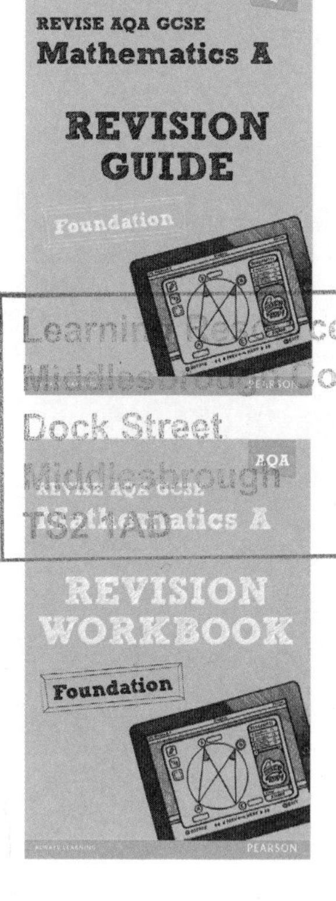

THE REVISE AQA SERIES

Available in print or online

Online editions for all titles in the Revise AQA series are available Summer 2013.

Presented on our ActiveLearn platform, you can view the full book and customise it by adding notes, comments and weblinks.

Print editions

Mathematics A Revision Guide Foundation 9781447941323

Mathematics A Revision Workbook Foundation 9781447941408

Online editions

Mathematics A Revision Guide Foundation 9781447941330

Mathematics A Revision Workbook Foundation 9781447941415

This Revision Guide is designed to complement your classroom and home learning, and to help prepare you for the exam. It does not include all the content and skills needed for the complete course. It is designed to work in combination with Pearson's main AQA GCSE Mathematics 2010 Series.

To find out more visit:
www.pearsonschools.co.uk/aqagcsemathsrevision

Contents

1-to-1 page match with the Foundation Workbook ISBN 9781447941408

UNIT 1 STATISTICS AND NUMBER
1 Place value
2 Rounding numbers
3 Fractions
4 Calculator skills
5 Percentages
6 Percentage change 1
7 Ratio 1
8 The Data Handling Cycle
9 Collecting data
10 Surveys
11 Two-way tables
12 Pictograms
13 Bar charts
14 Measuring and drawing angles
15 Pie charts
16 Averages and range
17 Stem-and-leaf diagrams
18 Averages from tables 1
19 Averages from tables 2
20 Scatter graphs
21 Frequency polygons
22 Probability 1
23 Probability 2
24 Combinations
25 Relative frequency
26 Comparing data
27 Problem-solving practice 1
28 Problem-solving practice 2

UNIT 2 NUMBER AND ALGEBRA
29 Adding and subtracting
30 Multiplying and dividing
31 Decimals and place value
32 Operations on decimals
33 Decimals and estimation
34 Negative numbers
35 Squares, cubes and roots
36 Factors, multiples and primes
37 HCF and LCM
38 Operations on fractions
39 Mixed numbers
40 Fractions, decimals and percentages
41 Percentage change 2
42 Ratio 2
43 Collecting like terms
44 Simplifying expressions
45 Indices
46 Expanding brackets
47 Factorising
48 Equations 1
49 Equations 2
50 Number machines
51 Inequalities
52 Solving inequalities
53 Substitution
54 Formulae
55 Writing formulae
56 Rearranging formulae
57 Using algebra
58 Coordinates
59 Straight-line graphs 1
60 Straight-line graphs 2
61 Real-life graphs 1
62 Distance–time graphs
63 Sequences 1
64 Sequences 2
65 Problem-solving practice 1
66 Problem-solving practice 2

UNIT 3 GEOMETRY AND ALGEBRA
67 Proportion
68 Trial and improvement
69 Quadratic graphs
70 Using quadratic graphs
71 Real-life graphs 2
72 Measuring lines
73 Metric units
74 Angles 1
75 Angles 2
76 Solving angle problems
77 Angles in polygons
78 Symmetry
79 Quadrilaterals
80 Perimeter and area
81 Using area formulae
82 Solving area problems
83 Circles
84 Area of a circle
85 3-D shapes
86 Volume
87 Prisms
88 Volume of a cylinder
89 Plans and elevations
90 Bearings
91 Scale drawings and maps
92 Constructing bisectors
93 Constructing triangles
94 Loci
95 Speed
96 Measures
97 Congruent shapes
98 Translations
99 Reflections
100 Rotations
101 Enlargements
102 Similar shapes
103 Pythagoras' theorem
104 Problem-solving practice 1
105 Problem-solving practice 2

106 ANSWERS

A small bit of small print

AQA publishes Sample Assessment Material and the Specification on its website. That is the official content and this book should be used in conjunction with it. The questions in *Now try this* have been written to help you practise every topic in the book. Remember: the real exam questions may not look like this.

Target grades

Target grades are quoted in this book for some of the questions. Students targeting this grade should be aiming to get most of the marks available. Students targeting a higher grade should be aiming to get all of the marks available.

C D E F G

Place value

The value of each digit in a number depends on its position. Digits which are further to the left are worth more. You can use a place value diagram to help you read and write numbers.

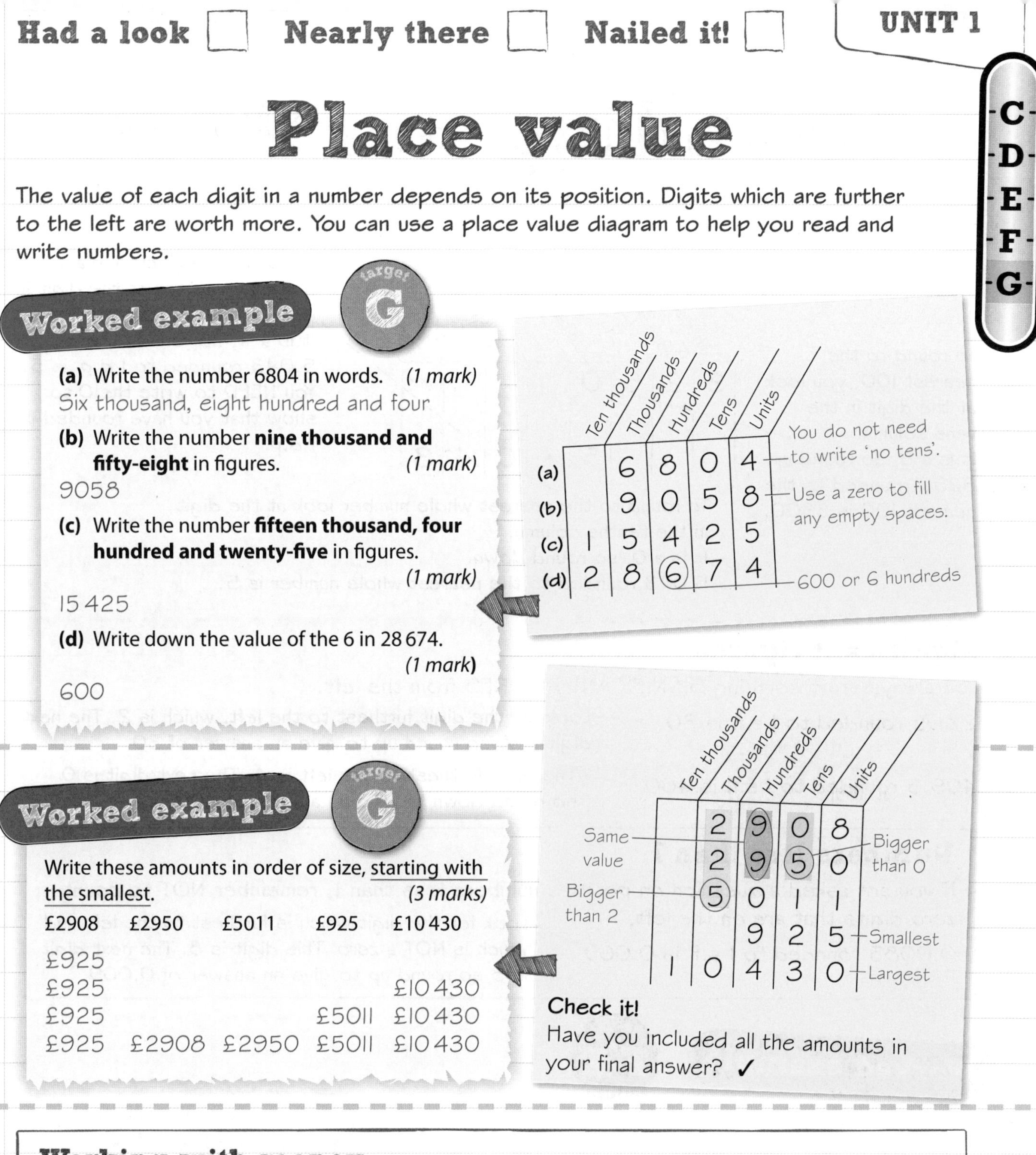

Worked example (target G)

(a) Write the number 6804 in words. *(1 mark)*

Six thousand, eight hundred and four

(b) Write the number **nine thousand and fifty-eight** in figures. *(1 mark)*

9058

(c) Write the number **fifteen thousand, four hundred and twenty-five** in figures. *(1 mark)*

15 425

(d) Write down the value of the 6 in 28 674. *(1 mark)*

600

Worked example (target G)

Write these amounts in order of size, starting with the smallest. *(3 marks)*

£2908 £2950 £5011 £925 £10 430

£925

£925 £10 430

£925 £5011 £10 430

£925 £2908 £2950 £5011 £10 430

Check it!

Have you included all the amounts in your final answer? ✓

Working with money

- Do all your calculations in the same units, either £ or p. ✓
- Write either £ or p in your answer, but not both. ✓
- 100 p = £1 ✓
- Amounts in pounds need 2 decimal places. Write 280 p as £2.80. ✓

Now try this (target G)

1 **(a)** Write the number sixteen thousand, three hundred and fifty-four in figures. *(1 mark)*
(b) Write the number 40 039 in words. *(1 mark)*
(c) Write down the value of the figure 8 in the number 27 586. *(1 mark)*

2 Write these amounts in order of size stating with the smallest. *(3 marks)*

£1974 £974 £1749 £1497 £1947

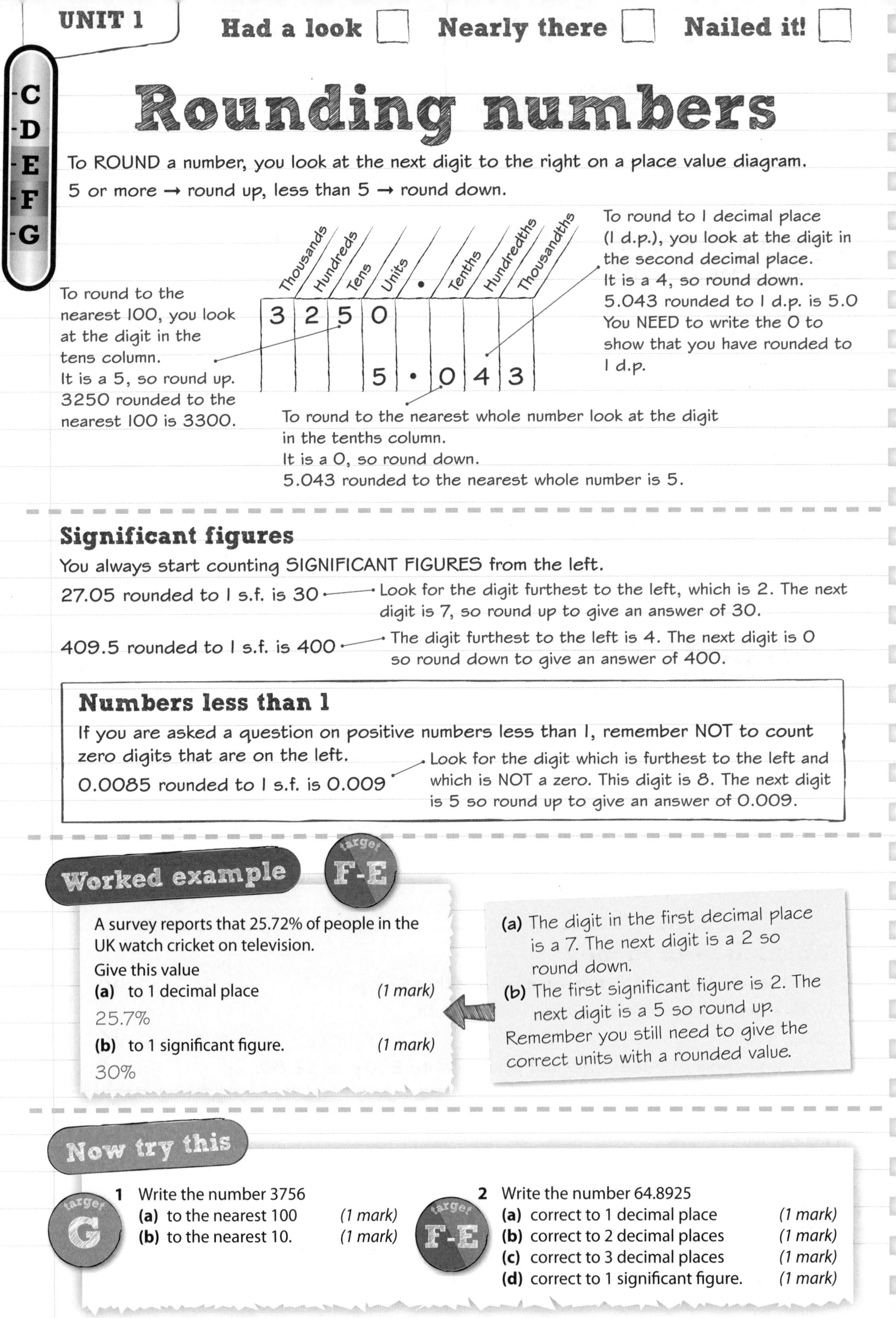

C D E F G

Rounding numbers

To ROUND a number, you look at the next digit to the right on a place value diagram.
5 or more → round up, less than 5 → round down.

Thousands	Hundreds	Tens	Units	•	Tenths	Hundredths	Thousandths
3	2	5	0				
			5	•	0	4	3

To round to the nearest 100, you look at the digit in the tens column.
It is a 5, so round up.
3250 rounded to the nearest 100 is 3300.

To round to 1 decimal place (1 d.p.), you look at the digit in the second decimal place.
It is a 4, so round down.
5.043 rounded to 1 d.p. is 5.0
You NEED to write the 0 to show that you have rounded to 1 d.p.

To round to the nearest whole number look at the digit in the tenths column.
It is a 0, so round down.
5.043 rounded to the nearest whole number is 5.

Significant figures

You always start counting SIGNIFICANT FIGURES from the left.

27.05 rounded to 1 s.f. is 30 — Look for the digit furthest to the left, which is 2. The next digit is 7, so round up to give an answer of 30.

409.5 rounded to 1 s.f. is 400 — The digit furthest to the left is 4. The next digit is 0 so round down to give an answer of 400.

Numbers less than 1

If you are asked a question on positive numbers less than 1, remember NOT to count zero digits that are on the left.

0.0085 rounded to 1 s.f. is 0.009 — Look for the digit which is furthest to the left and which is NOT a zero. This digit is 8. The next digit is 5 so round up to give an answer of 0.009.

Worked example

target F-E

A survey reports that 25.72% of people in the UK watch cricket on television.
Give this value

(a) to 1 decimal place *(1 mark)*

25.7%

(b) to 1 significant figure. *(1 mark)*

30%

(a) The digit in the first decimal place is a 7. The next digit is a 2 so round down.
(b) The first significant figure is 2. The next digit is a 5 so round up.
Remember you still need to give the correct units with a rounded value.

Now try this

target G

1 Write the number 3756

(a) to the nearest 100 *(1 mark)*
(b) to the nearest 10. *(1 mark)*

target F-E

2 Write the number 64.8925

(a) correct to 1 decimal place *(1 mark)*
(b) correct to 2 decimal places *(1 mark)*
(c) correct to 3 decimal places *(1 mark)*
(d) correct to 1 significant figure. *(1 mark)*

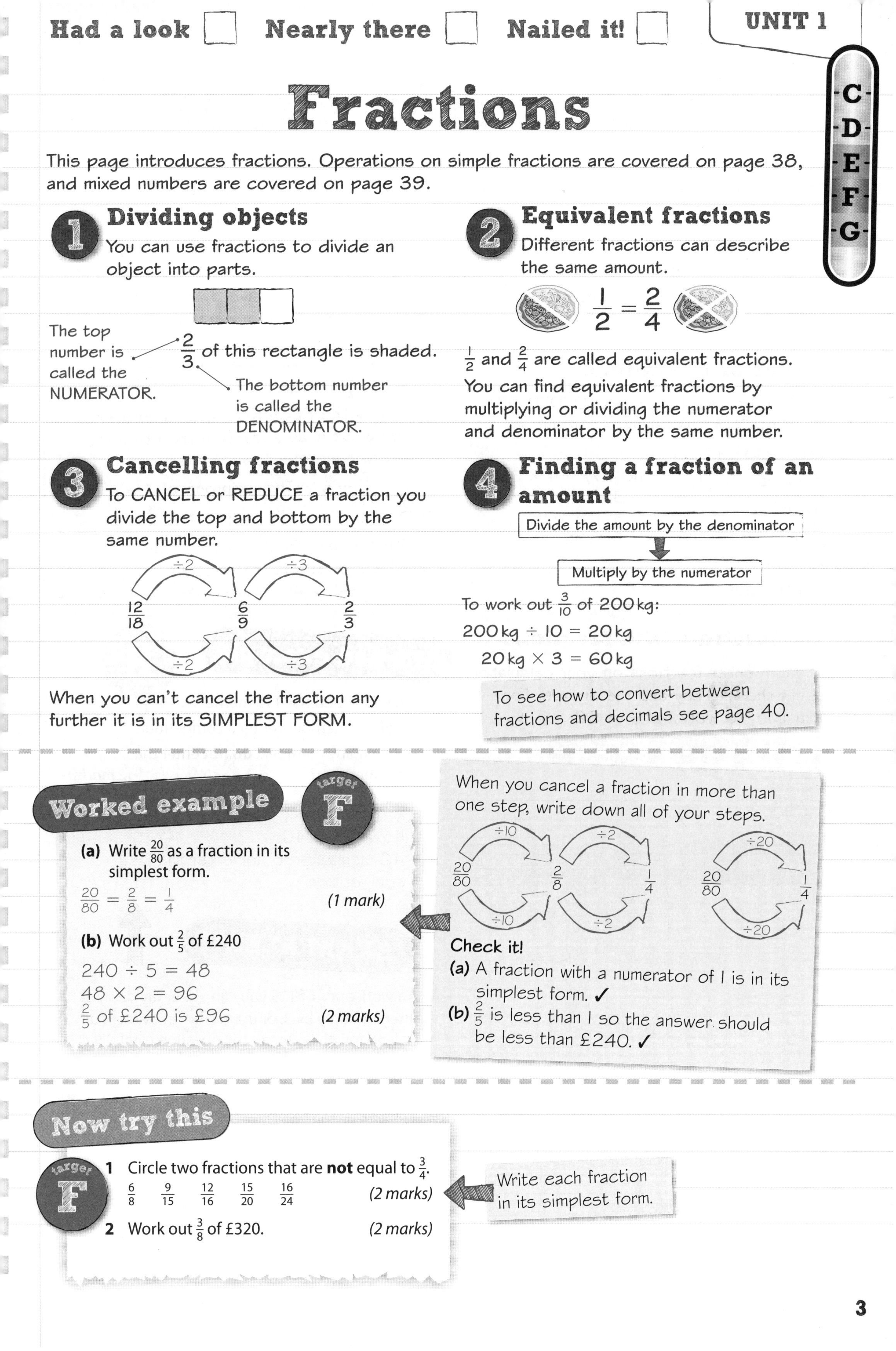

Fractions

This page introduces fractions. Operations on simple fractions are covered on page 38, and mixed numbers are covered on page 39.

1 Dividing objects

You can use fractions to divide an object into parts.

$\frac{2}{3}$ of this rectangle is shaded.

The top number is called the NUMERATOR.

The bottom number is called the DENOMINATOR.

2 Equivalent fractions

Different fractions can describe the same amount.

$\frac{1}{2} = \frac{2}{4}$

$\frac{1}{2}$ and $\frac{2}{4}$ are called equivalent fractions. You can find equivalent fractions by multiplying or dividing the numerator and denominator by the same number.

3 Cancelling fractions

To CANCEL or REDUCE a fraction you divide the top and bottom by the same number.

$\frac{12}{18} \xrightarrow{\div 2} \frac{6}{9} \xrightarrow{\div 3} \frac{2}{3}$

When you can't cancel the fraction any further it is in its SIMPLEST FORM.

4 Finding a fraction of an amount

Divide the amount by the denominator

↓

Multiply by the numerator

To work out $\frac{3}{10}$ of 200 kg:

200 kg ÷ 10 = 20 kg

20 kg × 3 = 60 kg

To see how to convert between fractions and decimals see page 40.

Worked example

Target F

(a) Write $\frac{20}{80}$ as a fraction in its simplest form.

$\frac{20}{80} = \frac{2}{8} = \frac{1}{4}$ *(1 mark)*

(b) Work out $\frac{2}{5}$ of £240

240 ÷ 5 = 48

48 × 2 = 96

$\frac{2}{5}$ of £240 is £96 *(2 marks)*

When you cancel a fraction in more than one step, write down all of your steps.

$\frac{20}{80} \xrightarrow{\div 10} \frac{2}{8} \xrightarrow{\div 2} \frac{1}{4}$

$\frac{20}{80} \xrightarrow{\div 20} \frac{1}{4}$

Check it!

(a) A fraction with a numerator of 1 is in its simplest form. ✓

(b) $\frac{2}{5}$ is less than 1 so the answer should be less than £240. ✓

Now try this

Target F

1 Circle two fractions that are **not** equal to $\frac{3}{4}$.

$\frac{6}{8}$ $\frac{9}{15}$ $\frac{12}{16}$ $\frac{15}{20}$ $\frac{16}{24}$ *(2 marks)*

Write each fraction in its simplest form.

2 Work out $\frac{3}{8}$ of £320. *(2 marks)*

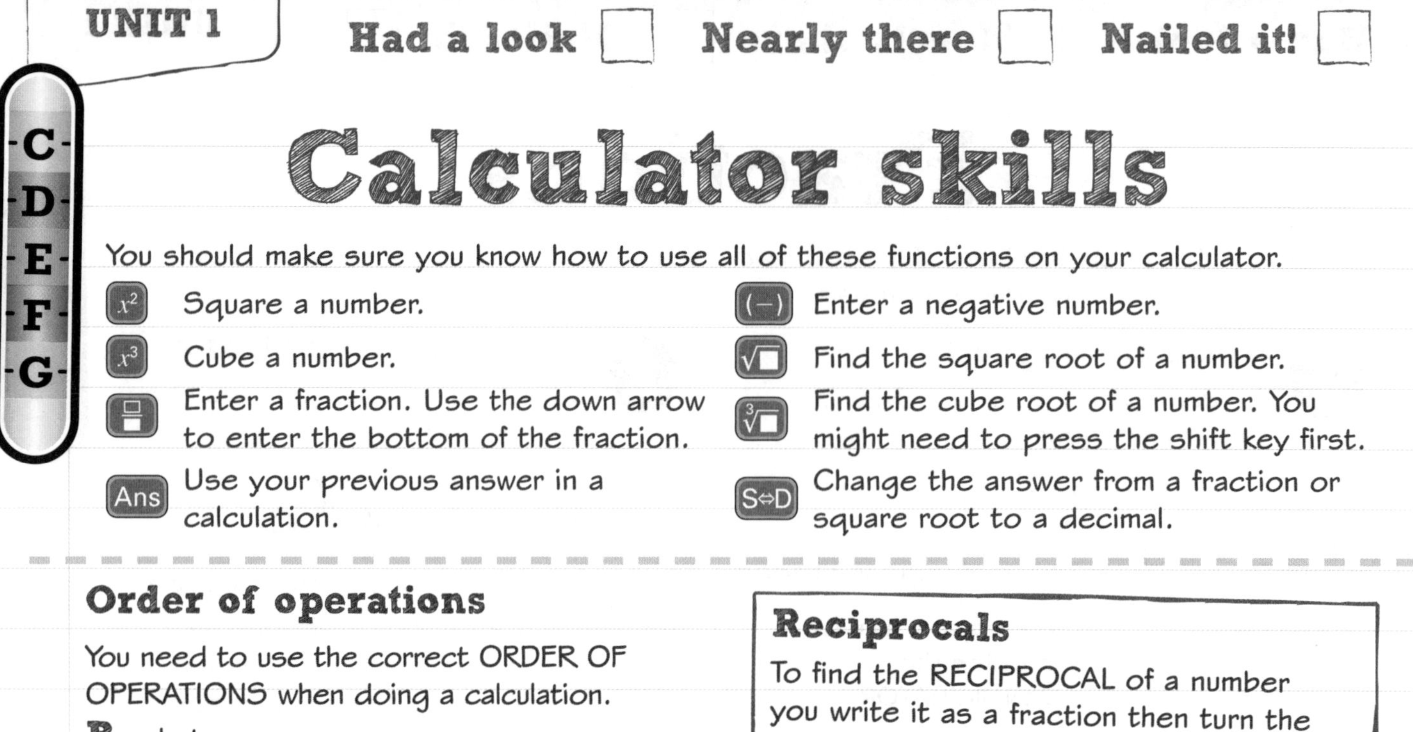

C D E F G

Calculator skills

You should make sure you know how to use all of these functions on your calculator.

- [x^2] Square a number.
- [x^3] Cube a number.
- [fraction key] Enter a fraction. Use the down arrow to enter the bottom of the fraction.
- [Ans] Use your previous answer in a calculation.
- [(−)] Enter a negative number.
- [$\sqrt{\square}$] Find the square root of a number.
- [$\sqrt[3]{\square}$] Find the cube root of a number. You might need to press the shift key first.
- [S⇔D] Change the answer from a fraction or square root to a decimal.

Order of operations

You need to use the correct ORDER OF OPERATIONS when doing a calculation.

Brackets
Indices
Division
Multiplication
Addition
Subtraction

$(10 - 7) + 4 \times 3^2$
$= 3 + 4 \times 3^2$
$= 3 + 4 \times 9$
$= 3 + 36$
$= 39$

Reciprocals

To find the RECIPROCAL of a number you write it as a fraction then turn the fraction upside down.

$7 = \frac{7}{1} \rightarrow \frac{1}{7}$ The reciprocal of 7 is $\frac{1}{7}$

$\frac{3}{4} \rightarrow \frac{4}{3}$ The reciprocal of $\frac{3}{4}$ is $\frac{4}{3}$

You can use the [x^{-1}] key on your calculator to find reciprocals.

Calculating with fractions

You can enter fractions on your calculator using the [fraction] key and the arrows. For example, to work out $\frac{3}{8}$ of 52:

[fraction] [3] [▼] [8] [▶] [×] [5] [2] [=]

Display: $\frac{3}{8} \times 52$ → $\frac{39}{2}$

If you want to convert an answer on your calculator display from a fraction to a decimal you can use the [S⇔D] key.

Worked example

Target E

A youth club has 115 members.
$\frac{3}{5}$ of the members enter a competition.
How many members **do not** enter the competition? *(2 marks)*

$\frac{3}{5} \times 115 = 69$
$115 - 69 = 46$
46 members do not enter the competition.

EXAM ALERT!

To work out $\frac{3}{5}$ of 115 you can either divide by 5 then multiply by 3, or use the [fraction] key on your calculator. Always read the question carefully. You need to subtract your value from 115 to get the final answer.

Students have struggled with exam questions similar to this – **be prepared!**

Now try this

Target E

1 Use your calculator to work out:

(a) 4.5^3 *(1 mark)*

(b) $1.7^2 + \sqrt{5.6}$ *(1 mark)*

Write down all the figures on your calculator display.

Target D

2 Use your calculator to work out

$$\frac{4.6 \times 3.8}{8.2 - 5.5}$$

Write down all the figures on your calculator display. *(2 marks)*

Write down what the top and bottom of the fraction come to before dividing.

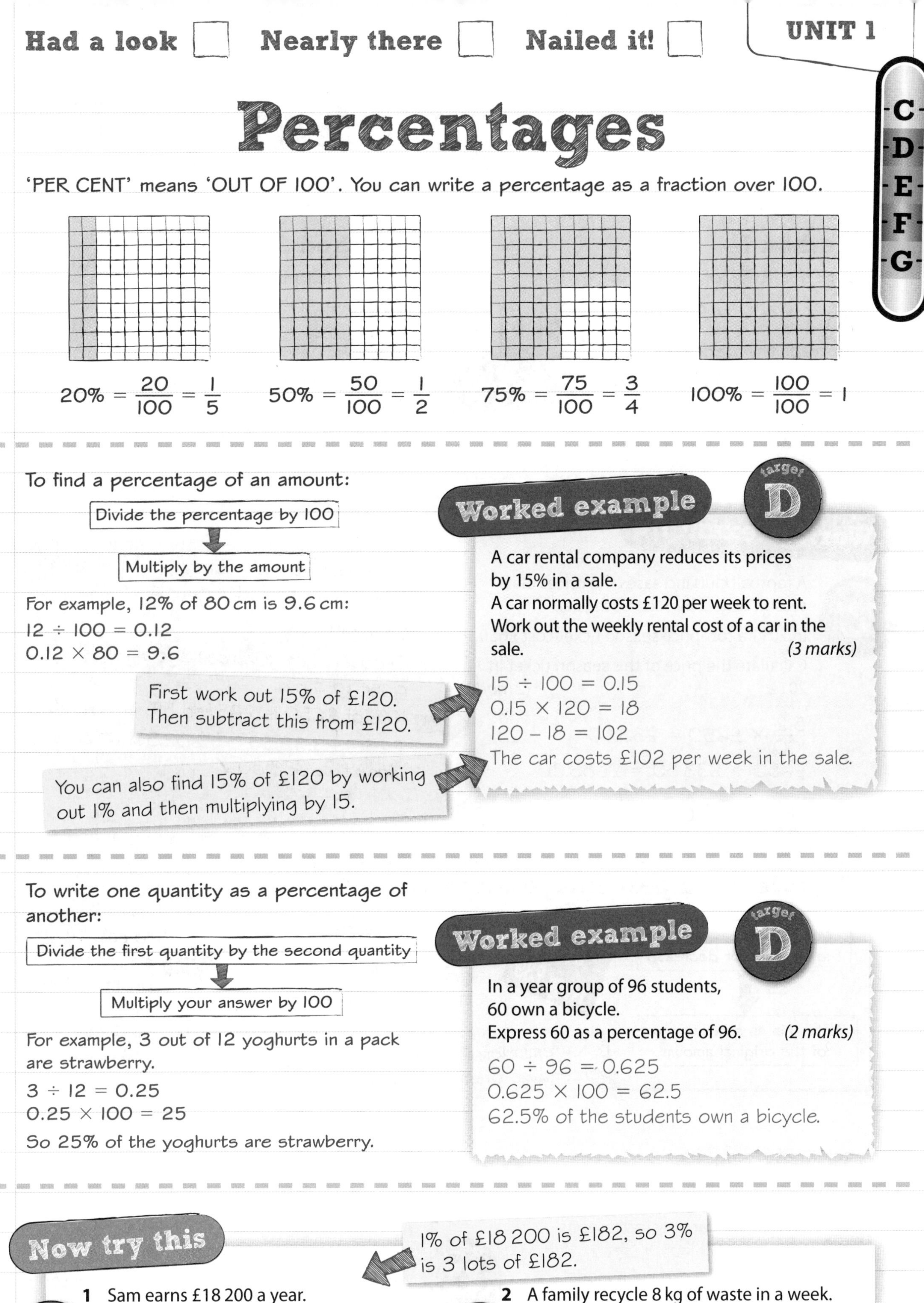

Percentages

'PER CENT' means 'OUT OF 100'. You can write a percentage as a fraction over 100.

$20\% = \frac{20}{100} = \frac{1}{5}$ $50\% = \frac{50}{100} = \frac{1}{2}$ $75\% = \frac{75}{100} = \frac{3}{4}$ $100\% = \frac{100}{100} = 1$

To find a percentage of an amount:

Divide the percentage by 100

↓

Multiply by the amount

For example, 12% of 80 cm is 9.6 cm:

$12 \div 100 = 0.12$

$0.12 \times 80 = 9.6$

Worked example

target D

A car rental company reduces its prices by 15% in a sale.
A car normally costs £120 per week to rent.
Work out the weekly rental cost of a car in the sale. *(3 marks)*

$15 \div 100 = 0.15$

$0.15 \times 120 = 18$

$120 - 18 = 102$

The car costs £102 per week in the sale.

First work out 15% of £120. Then subtract this from £120.

You can also find 15% of £120 by working out 1% and then multiplying by 15.

To write one quantity as a percentage of another:

Divide the first quantity by the second quantity

↓

Multiply your answer by 100

For example, 3 out of 12 yoghurts in a pack are strawberry.

$3 \div 12 = 0.25$

$0.25 \times 100 = 25$

So 25% of the yoghurts are strawberry.

Worked example

target D

In a year group of 96 students, 60 own a bicycle.
Express 60 as a percentage of 96. *(2 marks)*

$60 \div 96 = 0.625$

$0.625 \times 100 = 62.5$

62.5% of the students own a bicycle.

Now try this

1% of £18 200 is £182, so 3% is 3 lots of £182.

1 Sam earns £18 200 a year.
He is given a pay rise of 3%.
How much is his pay rise? *(2 marks)*

2 A family recycle 8 kg of waste in a week.
2.8 kg of this waste is paper.
What percentage of the recycled waste is paper? *(2 marks)*

Percentage change 1

There are two methods that can be used to increase or decrease an amount by a percentage.

Method 1

Work out 26% of £280:

$\frac{26}{100} \times £280 = £72.80$

Subtract the decrease:

£280 − £72.80 = £207.20

Method 2

Use a multiplier:

100% + 30% = 130%

$\frac{130}{100} = 1.3$

So the multiplier for a 30% increase is 1.3:

400 g × 1.3 = 520 g

Worked example

Target D

A football club increases the prices of its season tickets by 5.2% each year.

In 2011 a top-price season ticket cost £650.

Calculate the price of this season ticket in 2012. *(2 marks)*

$\frac{5.2}{100} \times £650 = £33.80$

£650 + £33.80 = £683.80

When working with money, you must give your answer to 2 decimal places.

Check it!

Choose an easy percentage which is close to 5.2%.

10% of £650 is £65, so 5% is £32.50.

£650 + £32.50 = £682.50, which is close to £683.80 ✓

Calculating a percentage increase or decrease

Work out the amount of the increase or decrease

↓

Write this as a percentage of the original amount

60 − 39 = 21

$\frac{21}{60} = 35\%$

This is a 35% decrease.

For a reminder about writing one quantity as a percentage of another, have a look at page 5.

A question may ask you to calculate a percentage **profit** or **loss** rather than an increase or decrease.

Now try this

1 A car manufacturer increases its prices by 8%.

The price of a particular model before the increase was £13 250.

What will this particular model cost after the price increase? *(3 marks)*

2 A TV originally cost £520.

In a sale it was priced at £340.

What was the percentage reduction in the price?

Give your answer to 1 decimal place. *(3 marks)*

Reduction means decrease. Work out the decrease as a percentage of the original price.

Ratio 1

RATIOS are used to compare quantities.

The ratio of apples to oranges is 3 : 2
The ratio 3 : 2 is in simplest form.

You can write a ratio as a fraction.
$\frac{2}{5}$ of the pieces of fruit are oranges.

3 + 2 = 5

The denominator is the sum of the parts in the ratio.

Equivalent ratios

You can find equivalent ratios by multiplying or dividing by the same number.

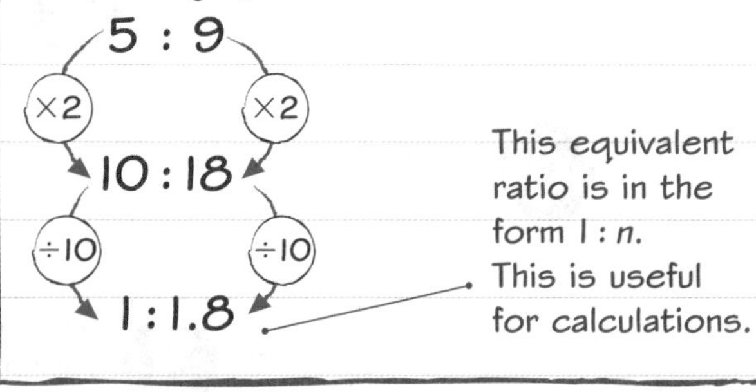

Jess and Simon are buying a new computer. They pay in the ratio 2 : 3.

Jess pays £194.

How much does Simon pay? *(3 marks)*

194 ÷ 2 = 97
3 × 97 = 291
Simon pays £291.

EXAM ALERT!

The order the names are written in is the same as the order of the numbers in the ratio. So 2 parts of the ratio represents the amount Jess paid. Divide £194 by 2 to work out how much each part of the ratio is worth, then multiply this value by 3 to find out how much Simon pays.

You can use a calculator, but remember to write down all your working.

Students have struggled with exam questions similar to this – **be prepared!**

A school has 200 students. 120 are female.
Work out the ratio of male to female students.
Give your answer in its simplest form. *(2 marks)*

200 – 120 = 80

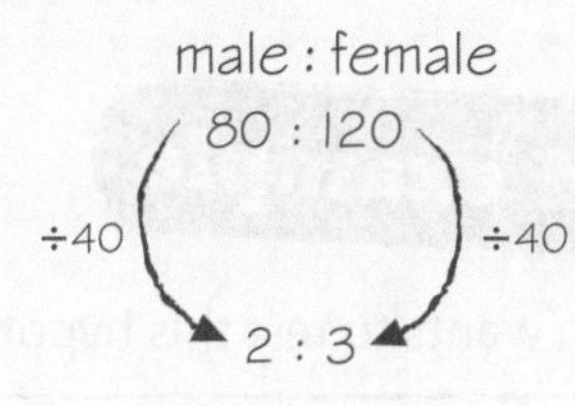

Simplest form

To write a ratio in its simplest form, find an equivalent ratio with the smallest possible whole number values.

Now try this

Start by calculating 330 ÷ 11 to work out how much each part of the ratio is worth.

1 A lemon and lime drink is made from lemonade and lime juice in the ratio 9 : 2.
How much lemonade will there be in 330 m*l* of the drink? *(3 marks)*

2 A bag contains red counters and green counters.
The ratio of red to green is 4 : 3.
There are 32 red counters in the bag.
What is the total number of counters in the bag? *(3 marks)*

C D E F G

The Data Handling Cycle

In your Unit 1 exam, you might have to write a plan for a statistical investigation.
A statistical investigation always follows the four components of the DATA HANDLING CYCLE.

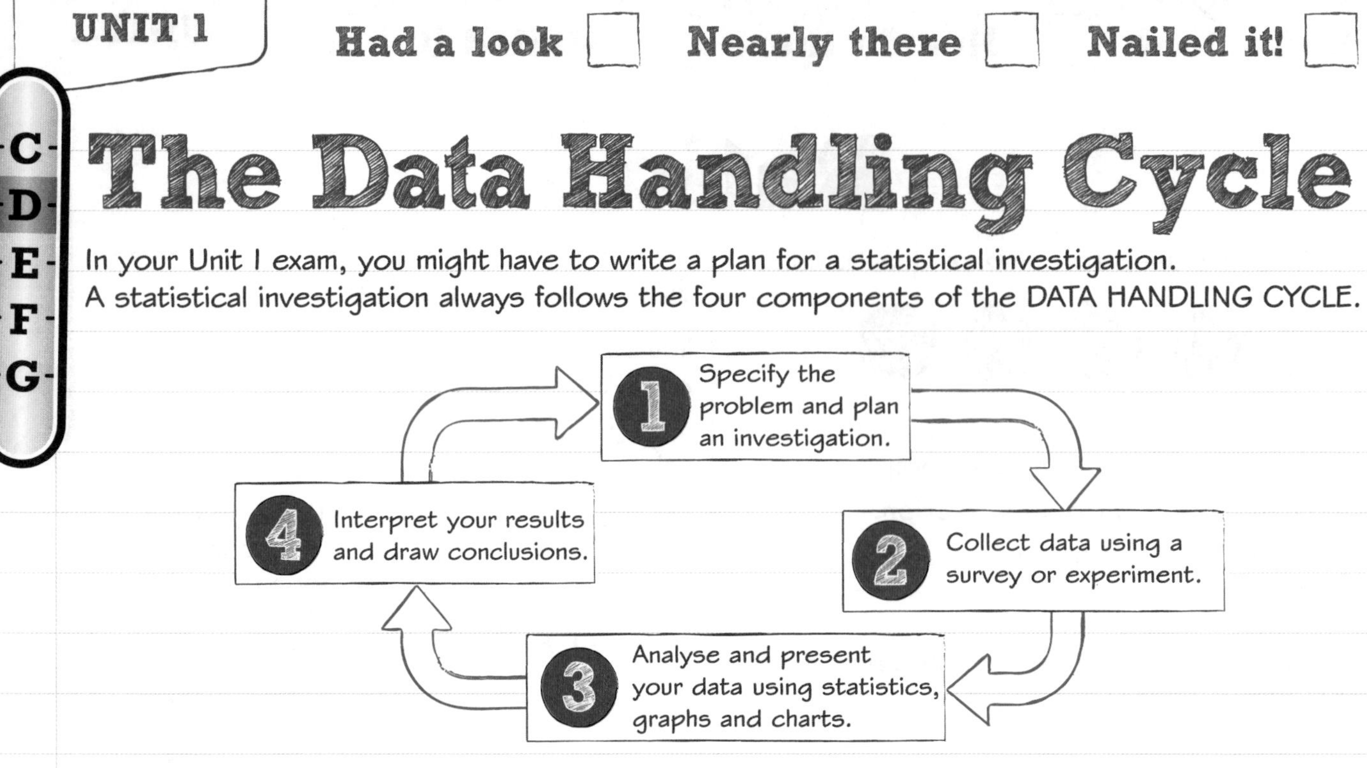

Hypothesis testing

In statistics, a hypothesis is a statement that might be either true or false. You can TEST whether the hypothesis is true by carrying out a statistical investigation.

Golden rule

When you're answering questions using the Data Handling Cycle, make sure your answers are specific to the hypothesis you want to test.

Worked example

target D

Sean wants to test this hypothesis:

Boys have larger handspans than girls.

Use the Data Handling Cycle to write a plan for Sean. *(3 marks)*

Collect data by measuring the handspans of 15 boys and 15 girls in my class to the nearest mm.

Calculate the mean and median handspan for the girls and the mean and median handspan for the boys.

Compare the averages for the boys and the girls and write a conclusion.

Use the four components of the Data Handling Cycle:

1. The problem has been given in the question so you don't need to write anything for this component.
2. Write down one way you could collect data such as a survey or an experiment.
3. Write down at least one way you could analyse or present your data, such as calculating statistics or drawing graphs.
4. Describe how you will use your statistics or graphs to interpret your results.

Now try this

target D

Sam wants to test this hypothesis:

Girls use social networking sites more often than boys.

Use the Data Handling Cycle to suggest how she should test this hypothesis. *(3 marks)*

Collecting data

You can collect PRIMARY data yourself, or get SECONDARY data from another source like the Internet. Data which is recorded in words, like the make of a car, is called QUALITATIVE data. Data which is recorded in numbers is called QUANTITATIVE data. Quantitative data can be DISCRETE, like the number of CDs you own, or CONTINUOUS, like time or length.

Methods of data collection

Here are examples of the main four methods of data collection:

1. OBSERVATION: Watching what foods students buy most often in the canteen.

2. EXPERIMENT: Measuring skin temperature before and after exercise.

3. SURVEY or QUESTIONNAIRE: Asking local people questions about their shopping habits.

4. DATA LOGGING: Using a sensor to count the number of cars that use a section of road.

Data logging is usually done with an electronic device.

Sampling

A SAMPLE is a small group chosen from a larger population.

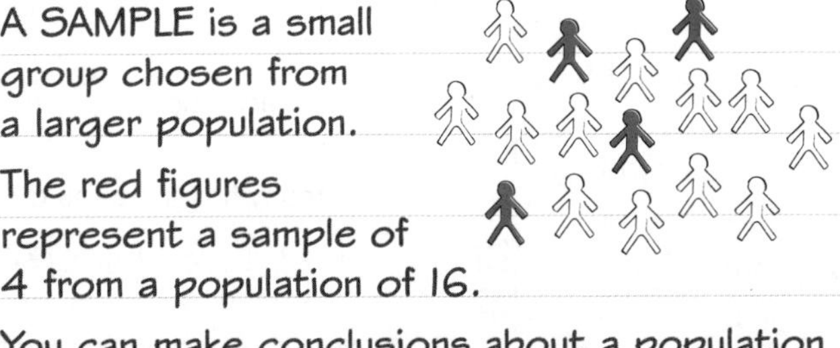

The red figures represent a sample of 4 from a population of 16.

You can make conclusions about a population by collecting data from a sample.

The larger your sample size, the more accurate your conclusions will be. One way to reduce BIAS in a statistical investigation is to use a RANDOM SAMPLE.

You can collect a random sample by numbering the population and selecting numbers at random using a computer.

Worked example

These two frequency tables show the same data.

A

Shoe size	7	$7\frac{1}{2}$	8	$8\frac{1}{2}$	9	$9\frac{1}{2}$	10	$10\frac{1}{2}$	11	$11\frac{1}{2}$	12
Frequency	2	4	5	4	11	5	7	7	4	3	1

B

Shoe size	$7-8\frac{1}{2}$	$9-10\frac{1}{2}$	$11-12\frac{1}{2}$
Frequency	15	30	8

(a) Give **one** advantage of table A over table B. *(1 mark)*

Table A shows the exact data that was collected.

(b) Give **one** advantage of table B over table A. *(1 mark)*

Table B shows a summary of the data which is easier to read.

Now try this

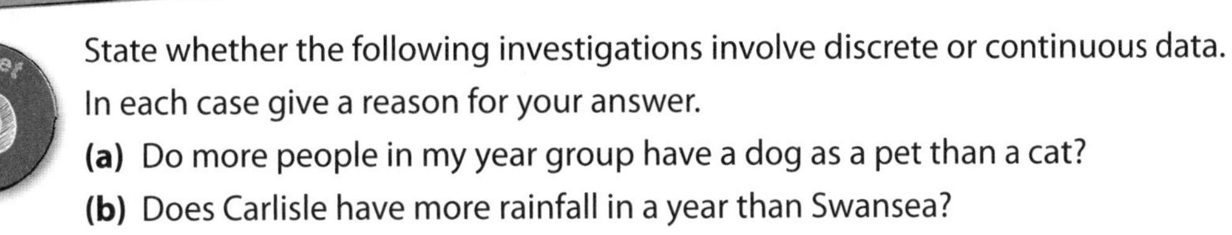

State whether the following investigations involve discrete or continuous data.
In each case give a reason for your answer.

(a) Do more people in my year group have a dog as a pet than a cat? *(2 marks)*

(b) Does Carlisle have more rainfall in a year than Swansea? *(2 marks)*

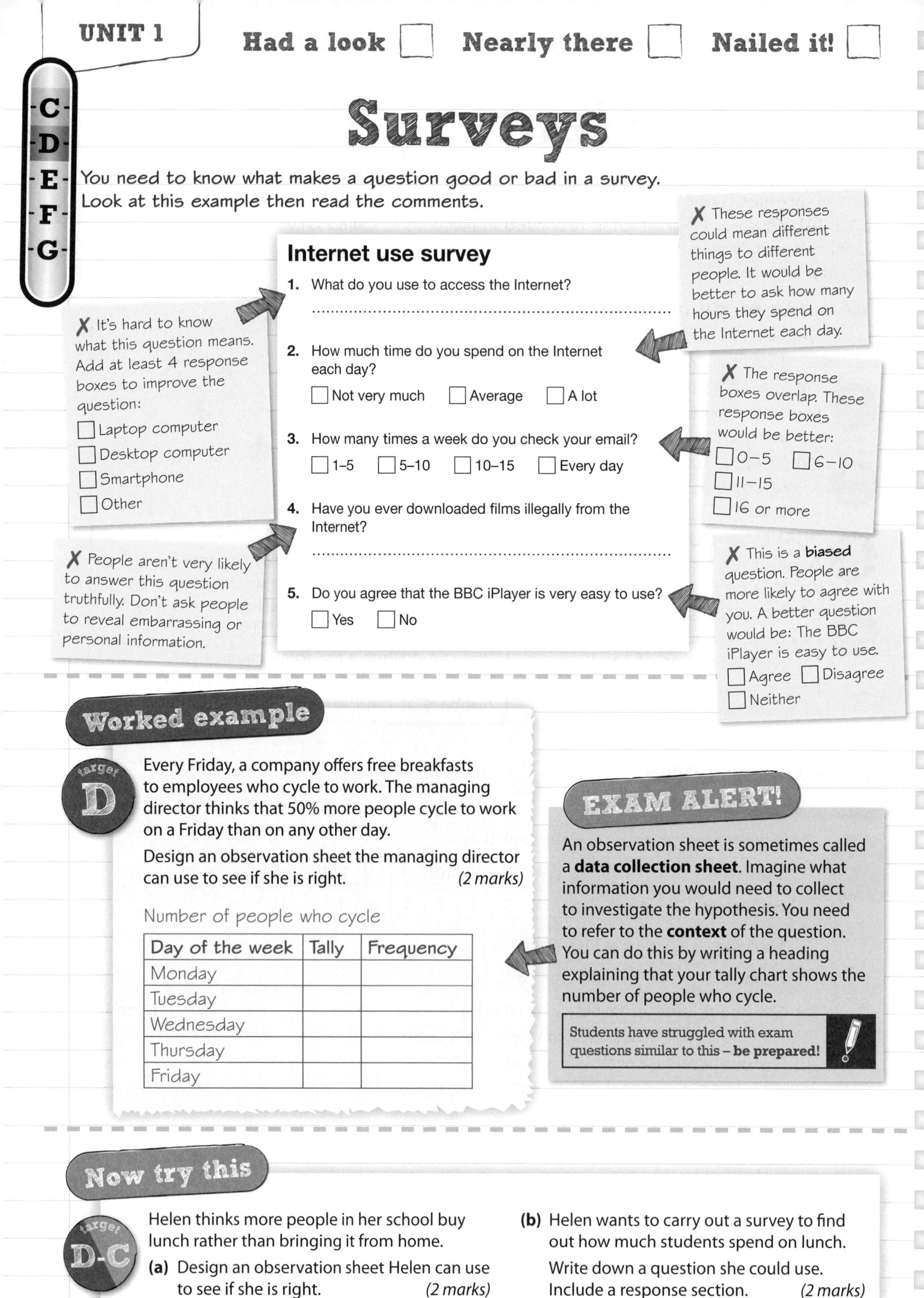

C D E F G

Surveys

You need to know what makes a question good or bad in a survey. Look at this example then read the comments.

Internet use survey

1. What do you use to access the Internet?

   ....

2. How much time do you spend on the Internet each day?

   ☐ Not very much ☐ Average ☐ A lot

3. How many times a week do you check your email?

   ☐ 1–5 ☐ 5–10 ☐ 10–15 ☐ Every day

4. Have you ever downloaded films illegally from the Internet?

   ....

5. Do you agree that the BBC iPlayer is very easy to use?

   ☐ Yes ☐ No

✗ It's hard to know what this question means. Add at least 4 response boxes to improve the question:
☐ Laptop computer
☐ Desktop computer
☐ Smartphone
☐ Other

✗ These responses could mean different things to different people. It would be better to ask how many hours they spend on the Internet each day.

✗ The response boxes overlap. These response boxes would be better:
☐ 0–5 ☐ 6–10
☐ 11–15
☐ 16 or more

✗ People aren't very likely to answer this question truthfully. Don't ask people to reveal embarrassing or personal information.

✗ This is a **biased** question. People are more likely to agree with you. A better question would be: The BBC iPlayer is easy to use.
☐ Agree ☐ Disagree
☐ Neither

Worked example

Target D

Every Friday, a company offers free breakfasts to employees who cycle to work. The managing director thinks that 50% more people cycle to work on a Friday than on any other day.

Design an observation sheet the managing director can use to see if she is right. *(2 marks)*

Number of people who cycle

Day of the week	Tally	Frequency
Monday		
Tuesday		
Wednesday		
Thursday		
Friday		

EXAM ALERT!

An observation sheet is sometimes called a **data collection sheet**. Imagine what information you would need to collect to investigate the hypothesis. You need to refer to the **context** of the question. You can do this by writing a heading explaining that your tally chart shows the number of people who cycle.

Students have struggled with exam questions similar to this – **be prepared!**

Now try this

Target D-C

Helen thinks more people in her school buy lunch rather than bringing it from home.

(a) Design an observation sheet Helen can use to see if she is right. *(2 marks)*

(b) Helen wants to carry out a survey to find out how much students spend on lunch.

Write down a question she could use. Include a response section. *(2 marks)*

C D E F G

Two-way tables

You can answer questions about two-way tables by adding or subtracting.

	Year 7	Year 8	Year 9	Total
Vegetarian	14	22	25	61
Not vegetarian	72	63	54	189
Total	86	85	79	250

There were 61 vegetarians in total.

In total 250 students were surveyed.

There were 86 Year 7 students surveyed.

There were 63 non-vegetarians in Year 8.

Worked example

Target D

Anton surveyed 120 people about how they voted at the last general election. He recorded the results in a two-way table:

	Labour	Conservative	Other	Total
Female	21	13	13	47
Male	32	27	14	73
Total	53	40	27	120

Complete the two-way table. *(4 marks)*

Labour column: $53 - 21 = 32$

Female row: $47 - 21 - 13 = 13$

Conservative column: $13 + 27 = 40$

Total row: $120 - 53 - 40 = 27$

Other column: $27 - 13 = 14$

Male row: $32 + 27 + 14 = 73$

Check: $47 + 73 = 120$

$53 + 40 + 27 = 120$ ✓

Everything in red is part of the answer.

Golden rules

The numbers in each column add up to the total for that column.

Other
13
+ 14
= 27

The numbers in each row add up to the total for that row.

Female	21	+ 13	+ 13	= 47

To complete a two-way table:

- write the total in the bottom right-hand cell
- look for rows and columns with only one missing number
- use addition and subtraction to find any missing values
- fill in the missing values as you go along.

Check it!

Add up the row totals and the column totals. They should be the same.

Now try this

Here is some information about class 9H.

There are 32 students altogether.

There are 4 more girls than boys.

A third of the girls are left-handed.

There are 23 right-handed students altogether.

Copy and complete the two-way table to show this information.

	Boys	Girls	Total
Left-handed			
Right-handed			
Total			32

(3 marks)

C D E F G

Pictograms

A PICTOGRAM can be used to represent data from a tally chart or frequency table.

This pictogram shows the results of a survey about how people watch television.

There is one row for each option.

Key: [TV] represents 2 people — A pictogram must have a KEY. This tells you how many items are represented by each picture.

Freeview	[TV] [TV] [TV] [TV] [TV] [TV]
Satellite	[TV] [TV] [TV] [TV] [½ TV]
Cable	[TV]
Internet	[TV] [TV] [TV] [½ TV]

[TV] represents 2 people so 12 people said they watched television using Freeview.

Each television represents 2 people, so half a television represents 1 person. This row represents 9 people.

To work out the total number of people in the survey, add together the totals of each row: 12 + 9 + 2 + 7 = 30.

Worked example

Target G

The pictogram shows the numbers of films Adam watched in June, July and August.

June	[4] [4]
July	[4] [3]
August	[4] [1]
September	[4] [4] [2]

Everything in red is part of the answer.

Key: [4 squares] represents 4 films

(a) Write down the number of films Adam watched in June. *(1 mark)*

8

(b) Write down the number of films Adam watched in August. *(1 mark)*

5

In September Adam watched 10 films.

(c) Use this information to complete the pictogram. *(1 mark)*

Use the key to work out what each picture represents.

[4 squares] = 4 films [3 squares] = 3 films

[2 squares] = 2 films [1 square] = 1 film

There is a block of 4 squares and a block of 1 square in August. This represents 4 + 1 = 5 films.

To represent 10 films you need two blocks of 4 squares and one block of 2 squares. Draw the squares neatly on the pictogram.

Tallies

A TALLY is a bit like a pictogram:

| represents 1 卌 represents 5

So to represent 12 you would draw:

卌 卌 ||

Now try this

A group of children were asked, 'What is your favourite type of TV programme?'

The pictogram shows some of the results.

(a) How many children chose Soaps? *(1 mark)*

(b) Half of the children chose either Soaps or Cartoons. How many children were in the group in total? *(2 marks)*

(c) Complete the pictogram to show how many children chose Sport. *(2 marks)*

Soaps	[4] [4] [1]
Cartoons	[4] [2]
Films	[3]
Sport	
Other	[4] [1]

Key: [4 squares] represents 4 children

Bar charts

You can use a BAR CHART to represent data given in a tally chart or frequency table. This DUAL BAR CHART shows the numbers of pairs of jeans owned by the members of a class.

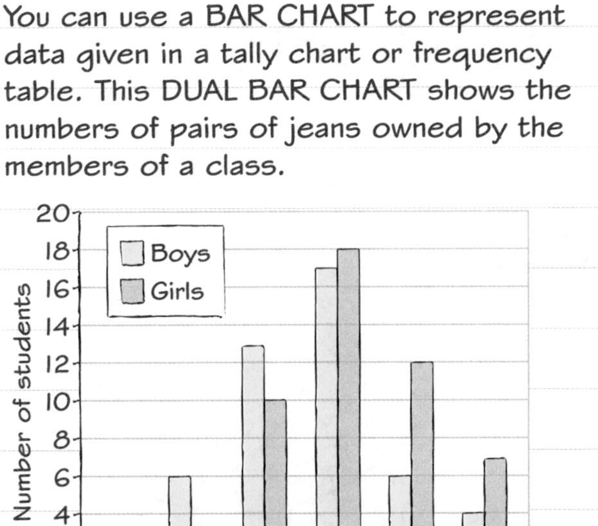

Bar chart features

- Bars are the same width. ✓
- There is a gap between the bars. ✓
- Both axes have labels. ✓
- Bars can be drawn horizontally or vertically. ✓
- The height (or length) of each bar represents the frequency. ✓
- In a dual bar chart two (or more) bars are drawn side by side. They can be used to compare data. ✓

Worked example

Kaitlyn carried out a survey of the colours of cars which passed the school gate in 10 minutes. Here are her results.

Colour	Tally
Red	𝍸 ||||
Black	𝍸 ||
Silver	𝍸 𝍸
Blue	|||

Use the grid to draw a <u>suitable chart or diagram</u> to represent Kaitlyn's results. *(4 marks)*

The question says 'suitable chart or diagram'. You could also represent this data using a pictogram or pie chart. To find the total number of cars from the bar chart, add up the values of each bar: 3 + 10 + 7 + 9 = 29.

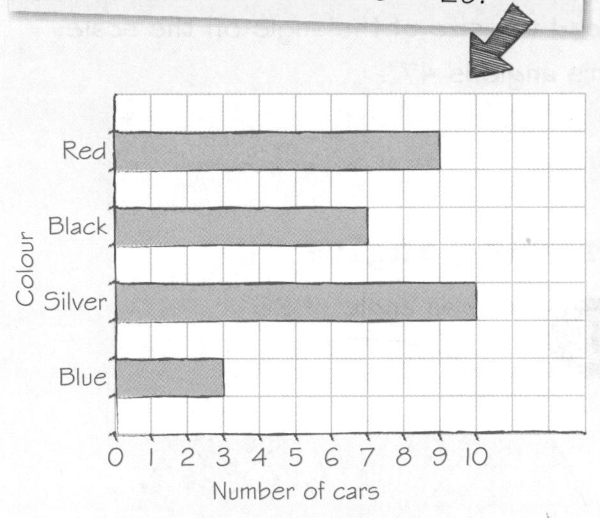

Now try this

target F

28 boys and 28 girls were asked to vote for their favourite subject from a list of English, Maths, Science and PE. The dual bar chart shows the results.

(a) Which subject was more popular with girls than boys? *(1 mark)*

(b) How many boys voted for either Maths or PE? *(1 mark)*

(c) Overall, which subject was the most popular? *(1 mark)*

(d) Which subject got one-quarter of all the votes? *(1 mark)*

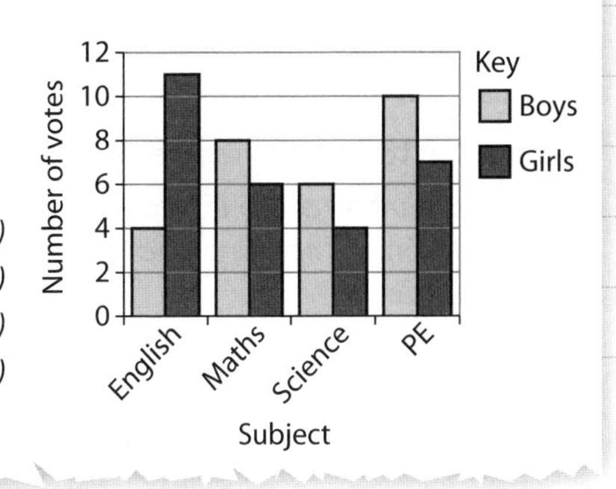

C D E F G

Measuring and drawing angles

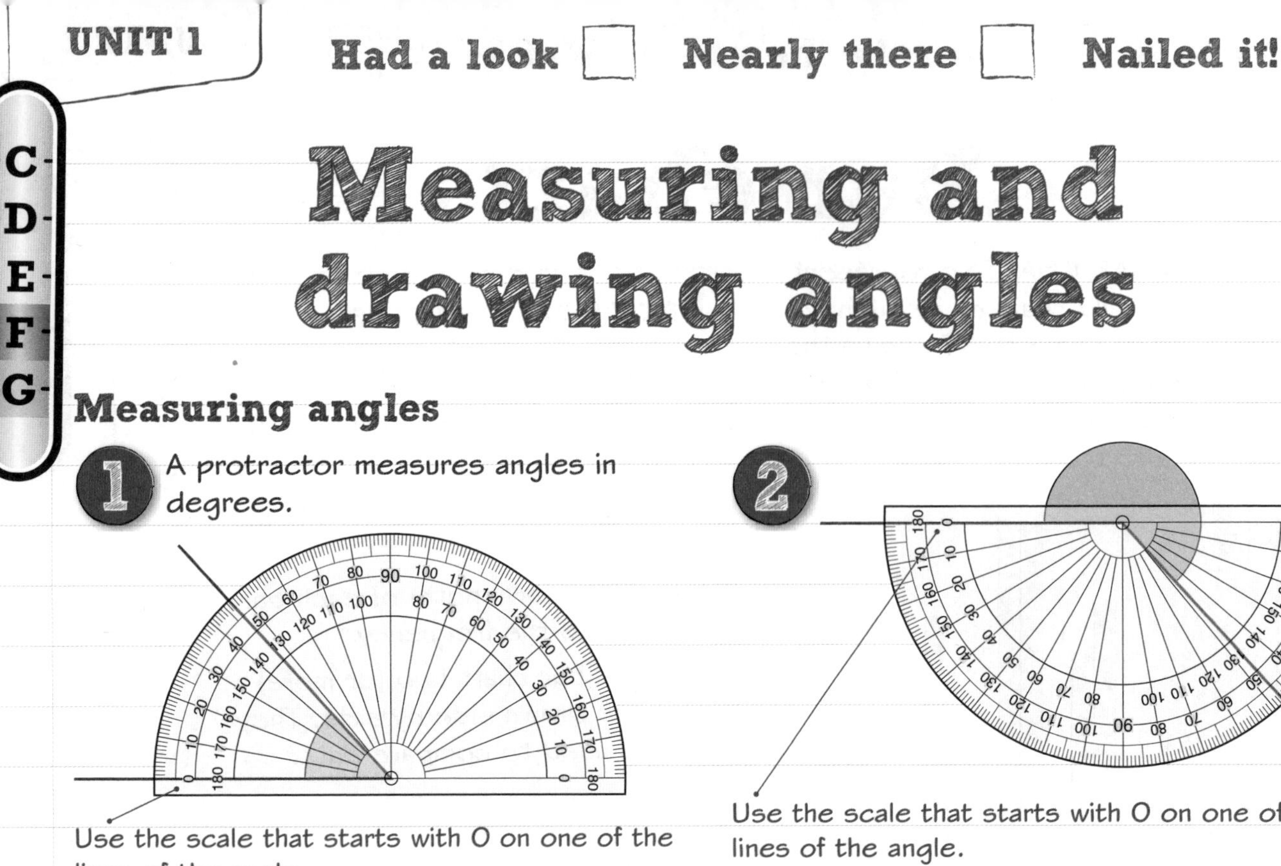

Measuring angles

1 A protractor measures angles in degrees.

Use the scale that starts with 0 on one of the lines of the angle.

Here, use the outside scale.

Place the centre of the protractor on the point of the angle.

Line up the zero line with one line of the angle.

Read the size of the angle off the scale.

This angle is 47°.

2 Use the scale that starts with 0 on one of the lines of the angle.

Here, use the inside scale.

To measure an angle bigger than 180° measure the smaller angle then subtract the answer from 360°.

$360 - 133 = 227$

The marked angle is 227°.

Estimate the size of an angle before measuring it. This lets you check that your answer is sensible.

Drawing angles

1 Draw an angle of 23°.

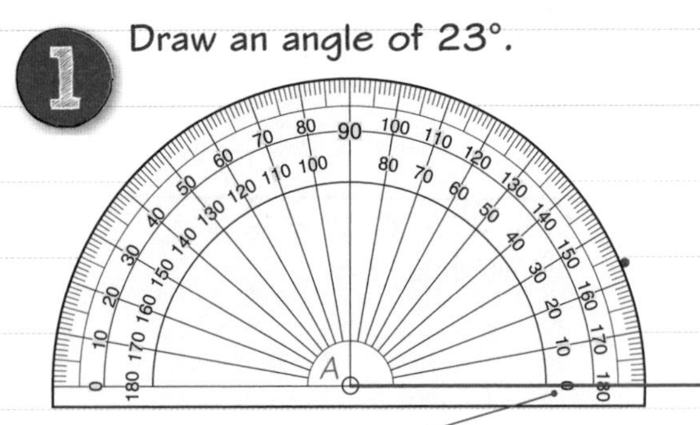

Use the scale that starts with 0 on one of the lines of the angle.

Here, use the inside scale.

Use a ruler to draw one line of the angle, *AB*.

Place the centre of your protractor on one end of the line. The zero line needs to lie along your line.

Find 23° on the scale. Draw a dot to mark this point.

2

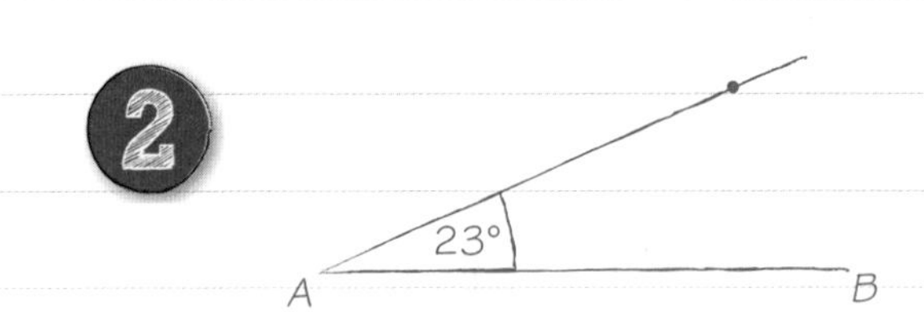

Use a ruler to join the end of the line and your dot with a straight line.

Draw in the angle curve and label your angle 23°.

Now try this

Measure the size of each of these angles.

(a) *(1 mark)* **(b)** *(1 mark)*

target F

Pie charts

In your exam you might have to interpret information given in a PIE CHART, or draw a pie chart from a frequency table. This pie chart shows the lunch choices made one day by 90 students.

There are 360° in a circle and 90 students.
$360 \div 90 = 4$
This means that each student is represented by an angle of 4°.

$64 \div 4 = 16$
So 64° represents 16 students.

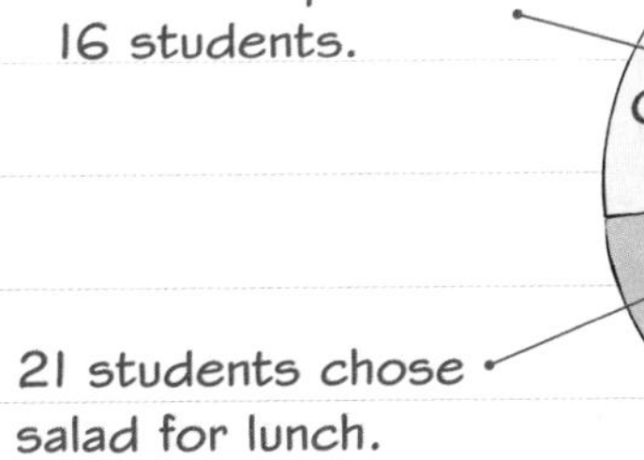

Half the students chose pizza. This SECTOR of the pie chart represents 45 students.

21 students chose salad for lunch.

Worked example

A farm has 40 fruit trees.
The table shows the number of each type of tree.
Draw a pie chart to represent this information.

Type of fruit tree	Number of trees	Angle
Apple	12	$12 \times 9° = 108°$
Plum	5	$5 \times 9° = 45°$
Pear	14	$14 \times 9° = 126°$
Peach	9	$9 \times 9° = 81°$

Angle for 1 tree $= 360° \div 40 = 9°$
Check: $108° + 45° + 126° + 81° = 360°$ ✓

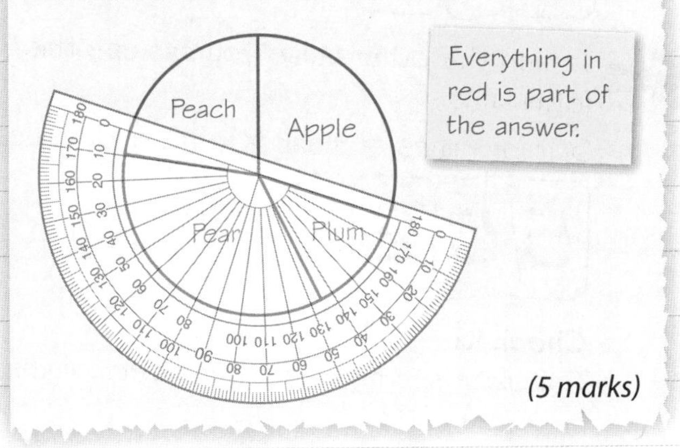

Everything in red is part of the answer.

(5 marks)

EXAM ALERT!

You need a sharp pencil, compasses and a protractor to draw a pie chart.

1. Add an 'Angle' column to the frequency table.
2. There are 360° in a full circle. There are 40 trees. So divide 360° by 40 to find the angle that represents 1 tree.
3. Multiply the angle that represents 1 tree by the number of each tree type to find the angle for each type.
4. **Check** that your angles add up to 360°.
5. Draw a circle using compasses. Draw a vertical line from the centre to the edge of the circle. Use a protractor to measure and draw the first angle (108°) from this line. Draw each angle carefully in order.
6. Label each sector of your pie chart with the type of fruit tree.

Students have struggled with exam questions similar to this – **be prepared!**

Now try this

There are 30 cars in a small car park.
The table shows the number of each make of car.
Draw a pie chart to represent this information. *(5 marks)*

Make of car	Number of cars
Audi	2
BMW	7
Ford	12
Nissan	4
Peugeot	5

30 cars = 360°
1 car = ...°

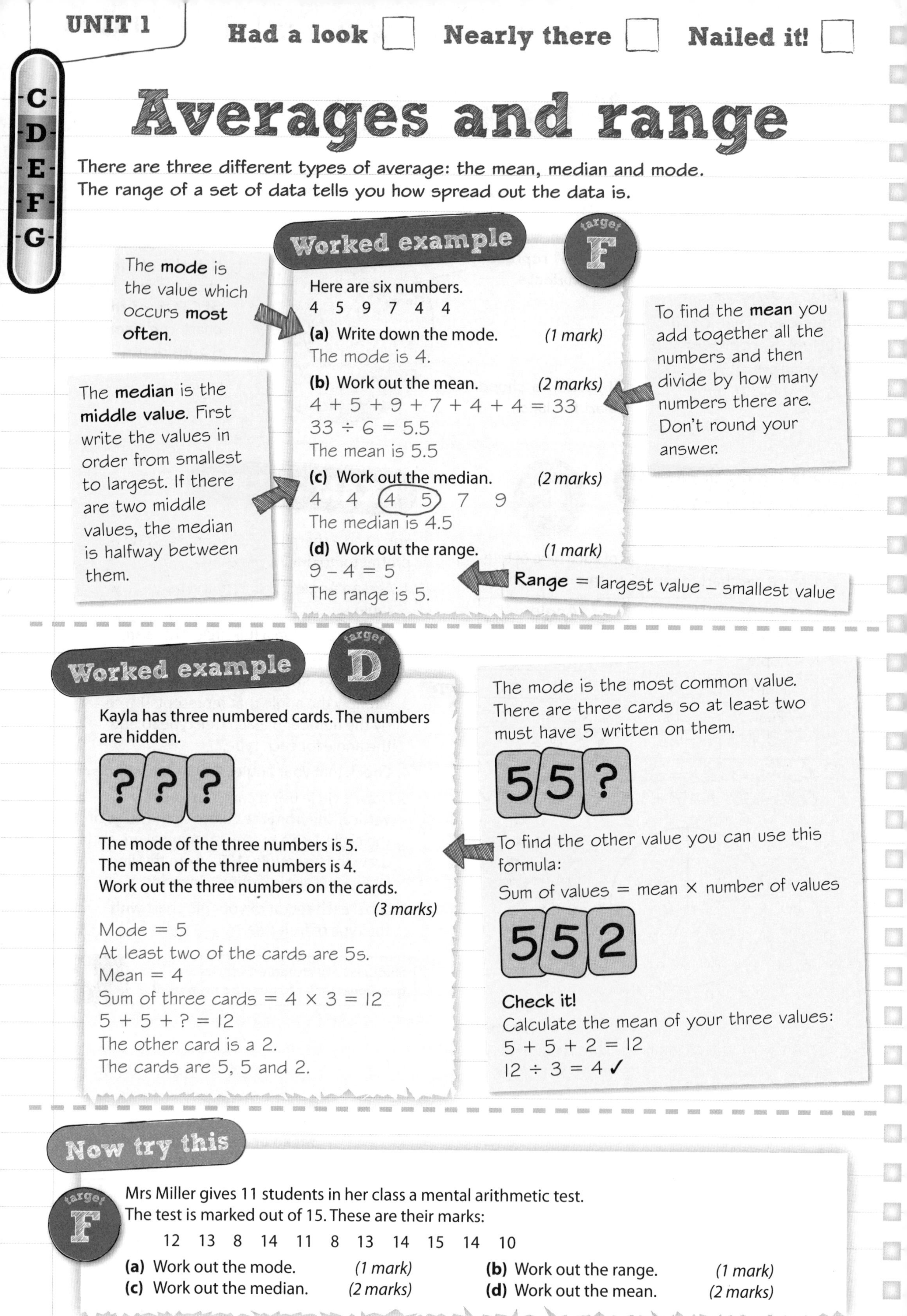

C D E F G

Averages and range

There are three different types of average: the mean, median and mode.
The range of a set of data tells you how spread out the data is.

Worked example (target F)

Here are six numbers.

4 5 9 7 4 4

(a) Write down the mode. *(1 mark)*

The mode is 4.

> The **mode** is the value which occurs **most often**.

(b) Work out the mean. *(2 marks)*

$4 + 5 + 9 + 7 + 4 + 4 = 33$

$33 \div 6 = 5.5$

The mean is 5.5

> To find the **mean** you add together all the numbers and then divide by how many numbers there are. Don't round your answer.

(c) Work out the median. *(2 marks)*

4 4 (4 5) 7 9

The median is 4.5

> The **median** is the **middle value**. First write the values in order from smallest to largest. If there are two middle values, the median is halfway between them.

(d) Work out the range. *(1 mark)*

$9 - 4 = 5$

The range is 5.

> **Range** = largest value − smallest value

Worked example (target D)

Kayla has three numbered cards. The numbers are hidden.

? ? ?

The mode of the three numbers is 5.
The mean of the three numbers is 4.
Work out the three numbers on the cards. *(3 marks)*

Mode = 5

At least two of the cards are 5s.

Mean = 4

Sum of three cards = $4 \times 3 = 12$

$5 + 5 + ? = 12$

The other card is a 2.

The cards are 5, 5 and 2.

> The mode is the most common value. There are three cards so at least two must have 5 written on them.
>
> 5 5 ?
>
> To find the other value you can use this formula:
>
> Sum of values = mean × number of values
>
> 5 5 2
>
> **Check it!**
>
> Calculate the mean of your three values:
>
> $5 + 5 + 2 = 12$
>
> $12 \div 3 = 4$ ✓

Now try this (target F)

Mrs Miller gives 11 students in her class a mental arithmetic test.
The test is marked out of 15. These are their marks:

12 13 8 14 11 8 13 14 15 14 10

(a) Work out the mode. *(1 mark)*
(b) Work out the range. *(1 mark)*
(c) Work out the median. *(2 marks)*
(d) Work out the mean. *(2 marks)*

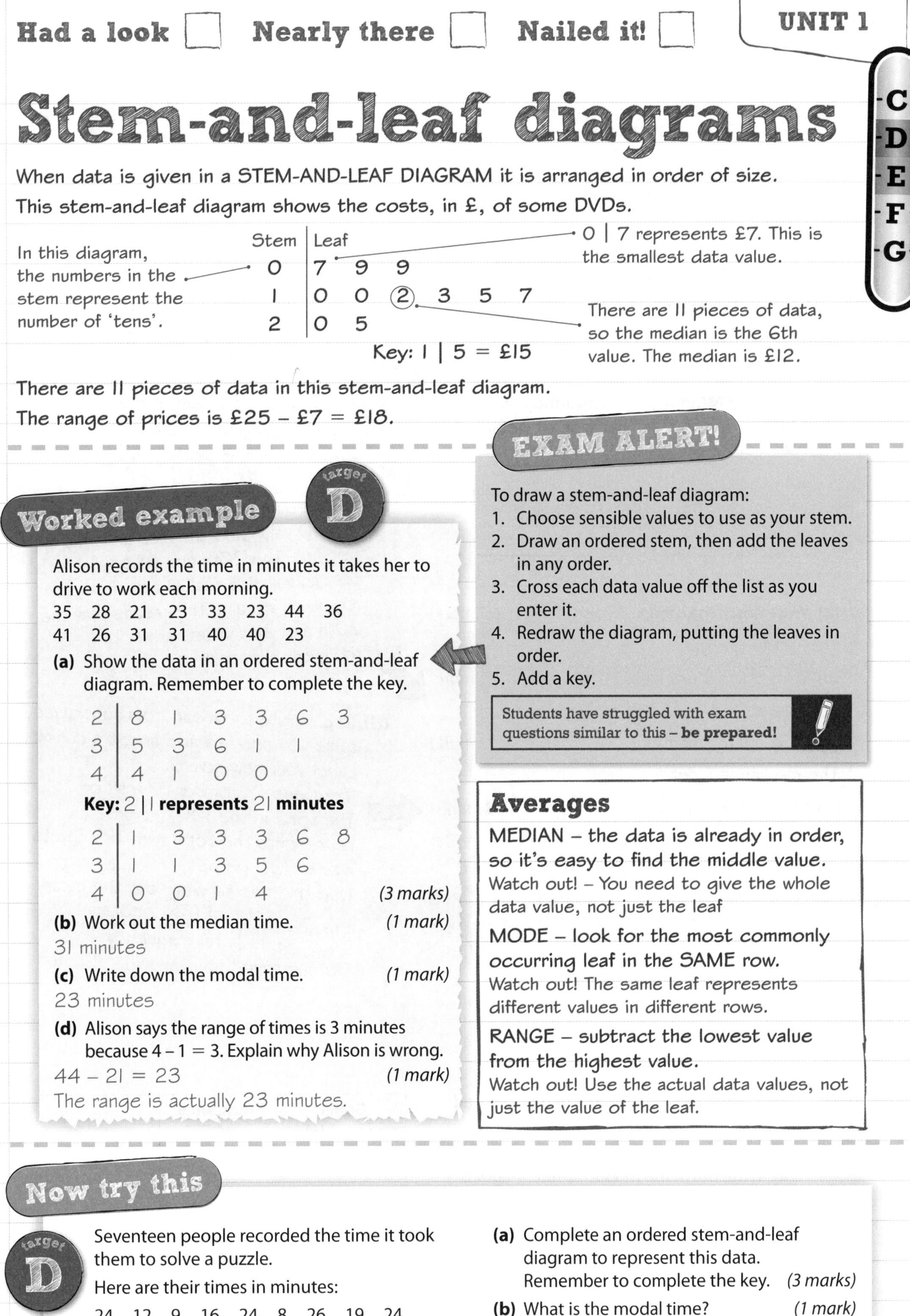

Stem-and-leaf diagrams

When data is given in a STEM-AND-LEAF DIAGRAM it is arranged in order of size.
This stem-and-leaf diagram shows the costs, in £, of some DVDs.

In this diagram, the numbers in the stem represent the number of 'tens'.

Stem	Leaf
0	7 9 9
1	0 0 (2) 3 5 7
2	0 5

Key: 1 | 5 = £15

0 | 7 represents £7. This is the smallest data value.

There are 11 pieces of data, so the median is the 6th value. The median is £12.

There are 11 pieces of data in this stem-and-leaf diagram.
The range of prices is £25 − £7 = £18.

Worked example

target D

Alison records the time in minutes it takes her to drive to work each morning.

35 28 21 23 33 23 44 36
41 26 31 31 40 40 23

(a) Show the data in an ordered stem-and-leaf diagram. Remember to complete the key.

2	8 1 3 3 6 3
3	5 3 6 1 1
4	4 1 0 0

Key: 2 | 1 **represents** 21 **minutes**

2	1 3 3 3 6 8
3	1 1 3 5 6
4	0 0 1 4

(3 marks)

(b) Work out the median time. *(1 mark)*

31 minutes

(c) Write down the modal time. *(1 mark)*

23 minutes

(d) Alison says the range of times is 3 minutes because 4 − 1 = 3. Explain why Alison is wrong. *(1 mark)*

44 − 21 = 23
The range is actually 23 minutes.

EXAM ALERT!

To draw a stem-and-leaf diagram:

1. Choose sensible values to use as your stem.
2. Draw an ordered stem, then add the leaves in any order.
3. Cross each data value off the list as you enter it.
4. Redraw the diagram, putting the leaves in order.
5. Add a key.

Students have struggled with exam questions similar to this – **be prepared!**

Averages

MEDIAN – the data is already in order, so it's easy to find the middle value.
Watch out! – You need to give the whole data value, not just the leaf

MODE – look for the most commonly occurring leaf in the SAME row.
Watch out! The same leaf represents different values in different rows.

RANGE – subtract the lowest value from the highest value.
Watch out! Use the actual data values, not just the value of the leaf.

Now try this

target D

Seventeen people recorded the time it took them to solve a puzzle.
Here are their times in minutes:

24 12 9 16 24 8 26 19 24
28 18 10 23 15 11 9 12

Use the actual data values, not just the values of the leaf.

(a) Complete an ordered stem-and-leaf diagram to represent this data. Remember to complete the key. *(3 marks)*

(b) What is the modal time? *(1 mark)*

(c) What is the median time? *(1 mark)*

(d) What is the range of the times? *(1 mark)*

C D E F G

Averages from tables 1

You need to be really careful when you are calculating averages from data given in a frequency table. Questions like this come up almost every year, so make sure you know the method.

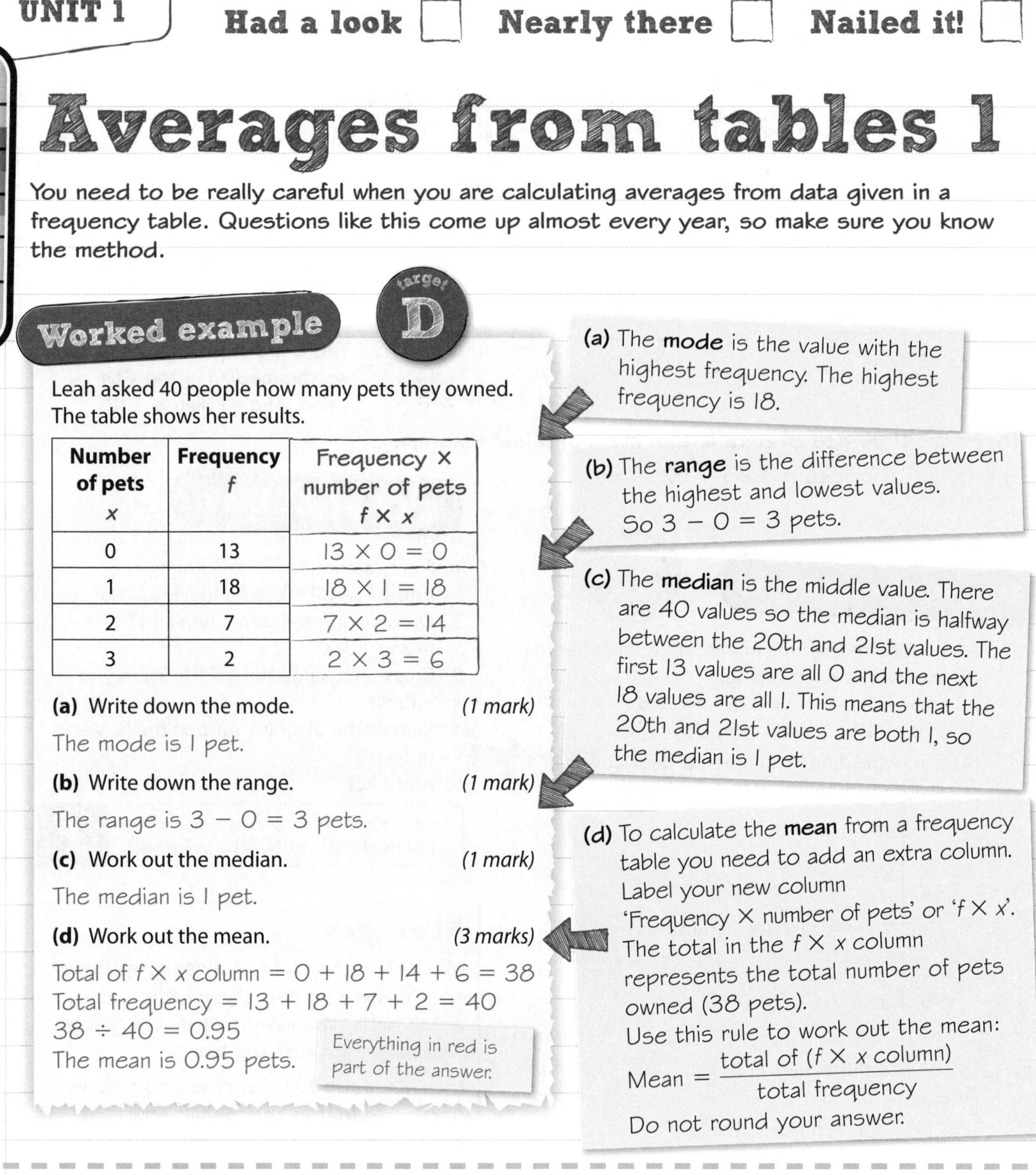

Worked example

target D

Leah asked 40 people how many pets they owned. The table shows her results.

Number of pets x	Frequency f	Frequency × number of pets $f \times x$
0	13	$13 \times 0 = 0$
1	18	$18 \times 1 = 18$
2	7	$7 \times 2 = 14$
3	2	$2 \times 3 = 6$

(a) Write down the mode. *(1 mark)*

The mode is 1 pet.

(b) Write down the range. *(1 mark)*

The range is $3 - 0 = 3$ pets.

(c) Work out the median. *(1 mark)*

The median is 1 pet.

(d) Work out the mean. *(3 marks)*

Total of $f \times x$ column $= 0 + 18 + 14 + 6 = 38$

Total frequency $= 13 + 18 + 7 + 2 = 40$

$38 \div 40 = 0.95$

The mean is 0.95 pets.

Everything in red is part of the answer.

(a) The **mode** is the value with the highest frequency. The highest frequency is 18.

(b) The **range** is the difference between the highest and lowest values. So $3 - 0 = 3$ pets.

(c) The **median** is the middle value. There are 40 values so the median is halfway between the 20th and 21st values. The first 13 values are all 0 and the next 18 values are all 1. This means that the 20th and 21st values are both 1, so the median is 1 pet.

(d) To calculate the **mean** from a frequency table you need to add an extra column. Label your new column 'Frequency × number of pets' or '$f \times x$'. The total in the $f \times x$ column represents the total number of pets owned (38 pets). Use this rule to work out the mean:

$$\text{Mean} = \frac{\text{total of } (f \times x \text{ column})}{\text{total frequency}}$$

Do not round your answer.

Now try this

The table shows the number of goals scored by 50 teams in matches taking place one weekend.

(a) Write down the modal number of goals scored. *(1 mark)*

(b) Work out the median number of goals scored. *(1 mark)*

(c) Work out the mean number of goals scored. *(3 marks)*

Goals scored (x)	Frequency (f)	
0	7	
1	15	
2	14	
3	7	
4	4	
5	3	

Use the extra column for 'Frequency × Goals scored ($f \times x$)'.

Averages from tables 2

C D E F G

Sometimes data in a frequency table is grouped into CLASS INTERVALS. You don't know the exact data values, but you can calculate an ESTIMATE of the mean, and write down which class interval contains the median and which one has the highest frequency.

Worked example

target C

Maisie recorded the times, in minutes, taken by 150 students to travel to school. The table shows her results.

Time (t minutes)	Frequency f	Midpoint x	$f \times x$
$0 \leqslant t < 20$	65	10	$65 \times 10 = 650$
$20 \leqslant t < 40$	42	30	$42 \times 30 = 1260$
$40 \leqslant t < 60$	39	50	$39 \times 50 = 1950$
$60 \leqslant t < 80$	4	70	$4 \times 70 = 280$

Everything in red is part of the answer.

(a) Write down the modal class interval. *(1 mark)*

$0 \leqslant t < 20$

(b) Write down the class interval which contains the median. *(1 mark)*

$20 \leqslant t < 40$

(c) Work out an estimate for the mean number of minutes that the students took to travel to school. *(4 marks)*

Sum of $f \times x$ column
$= 650 + 1260 + 1950 + 280 = 4140$

Total frequency
$= 65 + 42 + 39 + 4 = 150$

$4140 \div 150 = 27.6$ minutes.

Estimated mean $= 27.6$ minutes.

1. Add extra columns to the table.
2. Use the first extra column for 'midpoint' and work out the midpoint of each class interval.
3. Label 'frequency' f and 'midpoint' x.
4. Use the final column for $f \times x$. Make sure you use the midpoint when calculating $f \times x$ for each row.
5. Use this rule to estimate the mean.

$$\text{Estimate of mean} = \frac{\text{Total of } f \times x \text{ column}}{\text{Total frequency}}$$

(d) Explain why your answer to part **(c)** is an estimate. *(1 mark)*

Because you don't know the exact data values.

Now try this

The table shows the marks obtained by 50 students in a Maths test.

(a) Write down the modal class interval. *(1 mark)*

(b) Write down the class interval that contains the median. *(1 mark)*

(c) Work out an estimate of the mean mark for these students. *(4 marks)*

Marks (x)	Frequency (f)
$0 < x \leqslant 10$	10
$10 < x \leqslant 20$	11
$20 < x \leqslant 30$	8
$30 < x \leqslant 40$	6
$40 < x \leqslant 50$	10
$50 < x \leqslant 60$	5

C D E F G

Scatter graphs

The points on a scatter graph aren't always scattered. If the points are almost on a straight line then the scatter graph shows CORRELATION. The better the straight line, the stronger the correlation.

Negative correlation — Distance from London / House price

No correlation — Score on maths test / Weight

Positive correlation — Length of spring / Weight on spring

An ISOLATED point on a scatter graph is an extreme point that lies outside the normal range of values.

Worked example

target D

This scatter graph shows the engine capacity of some cars and the distance they will travel on one gallon of petrol.

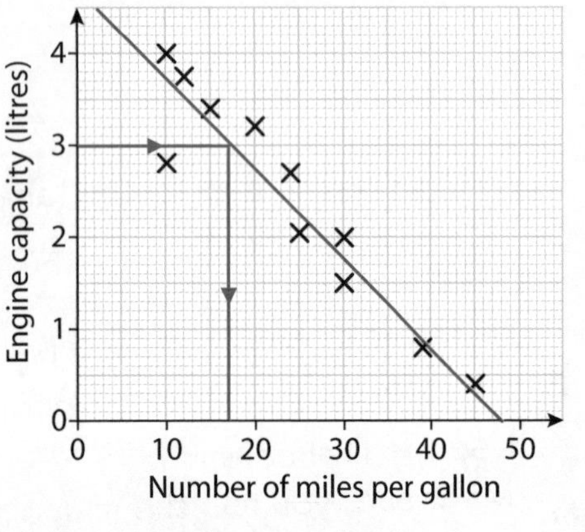

(a) Isaac has the hypothesis:

Cars with smaller engines are more efficient.

Comment on his hypothesis. *(1 mark)*

The graph shows negative correlation so there is strong support for his hypothesis.

(b) Estimate how far a car with a 3 litre engine will travel on 1 gallon of petrol. *(2 marks)*

17 miles

(a) There is negative correlation, so cars with bigger engines travel less far on one gallon of petrol. This supports Isaac's hypothesis. You can use words like 'strong' and 'weak' to show how much support the data provides for the hypothesis.

(b) Draw a line of best fit on the scatter graph. Read across from 3 on the vertical axis to the line of best fit, then down to the horizontal axis. Draw all the lines you use on the graph.

Line of best fit checklist

- Straight line that is as close as possible to all the points. ✓
- Used to predict values. ✓
- Does not need to go through (0, 0). ✓
- Drawn with a ruler. ✓
- Ignores isolated points. ✓

Now try this

target D

This scatter graph shows the daily hours of sunshine and the daily maximum temperature at 12 seaside resorts in England on one day last summer.

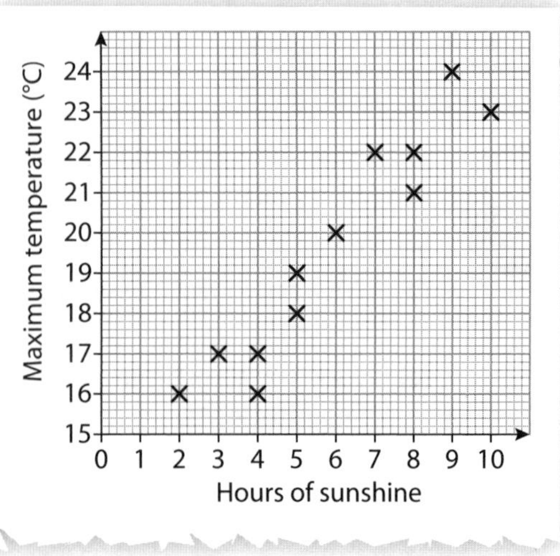

(a) Draw a line of best fit for this data. *(1 mark)*

(b) Is the correlation positive or negative? *(1 mark)*

(c) Describe the connection between the number of hours of sunshine and the maximum temperature. *(1 mark)*

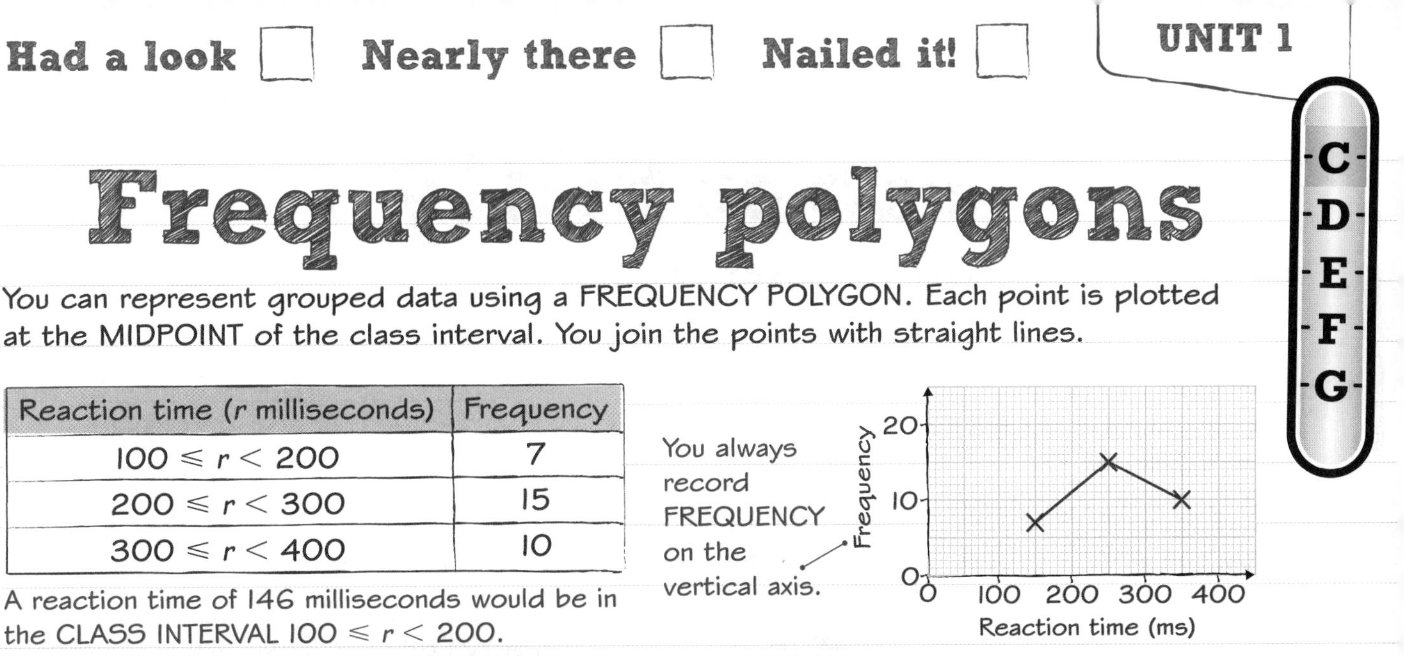

C D E F G

Frequency polygons

You can represent grouped data using a FREQUENCY POLYGON. Each point is plotted at the MIDPOINT of the class interval. You join the points with straight lines.

Reaction time (r milliseconds)	Frequency
$100 \leqslant r < 200$	7
$200 \leqslant r < 300$	15
$300 \leqslant r < 400$	10

A reaction time of 146 milliseconds would be in the CLASS INTERVAL $100 \leqslant r < 200$.

You always record FREQUENCY on the vertical axis.

Worked example

target C

30 students timed how long it took them to complete a jigsaw puzzle. The results were recorded in a grouped frequency table:

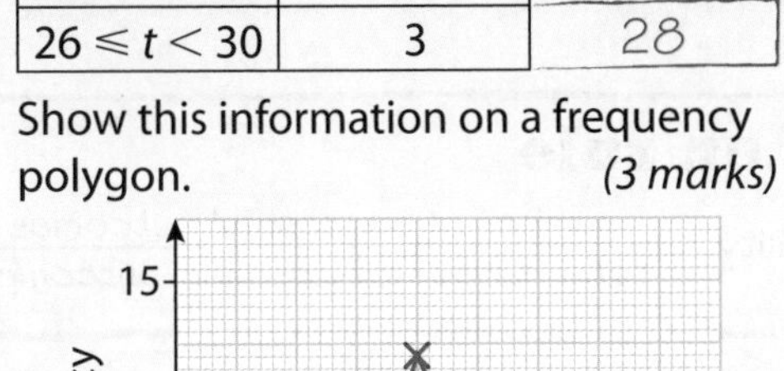

Time (t minutes)	Frequency	Midpoint
$10 \leqslant t < 14$	2	12
$14 \leqslant t < 18$	5	16
$18 \leqslant t < 22$	12	20
$22 \leqslant t < 26$	8	24
$26 \leqslant t < 30$	3	28

Show this information on a frequency polygon. *(3 marks)*

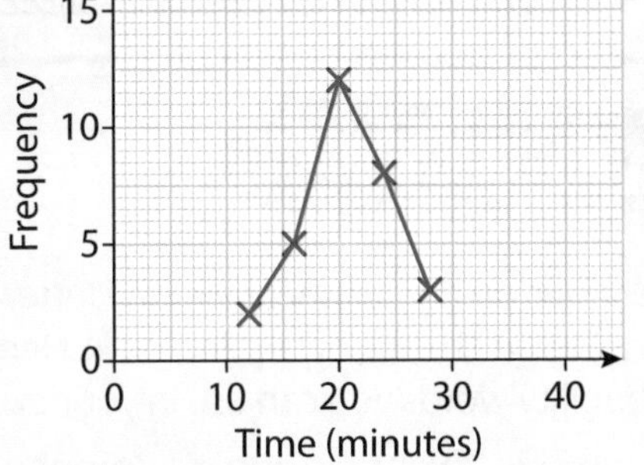

Start by working out the **midpoints** of the class intervals.
The midpoint of the class interval $10 \leqslant t < 14$ is $\frac{10 + 14}{2} = 12$

Check it!
In your exam you will only be asked to draw a frequency polygon for data with **equal class intervals**. So make sure that your midpoints are the same distance apart.

Estimating the mean

You can estimate the mean for data given in a frequency polygon using this formula:

$$\text{Mean} \approx \frac{\text{Sum of (midpoint} \times \text{frequency)}}{\text{Total frequency}}$$

For the worked example on the left:

$$\frac{(2 \times 12) + (5 \times 16) + (12 \times 20) + (8 \times 24) + (3 \times 28)}{30} = 20\frac{2}{3}$$

An estimate for the mean time is 20 minutes, 40 seconds.

There is more about estimating the mean of grouped data on page 19.

Now try this

This table shows the times taken, in minutes, for 50 people to solve a crossword puzzle.

(a) Draw a frequency polygon for this data. *(2 marks)*

(b) Work out an estimate for the mean time taken to solve the crossword puzzle. *(4 marks)*

Time t (minutes)	Frequency
$0 < t \leqslant 10$	3
$10 < t \leqslant 20$	9
$20 < t \leqslant 30$	11
$30 < t \leqslant 40$	18
$40 < t \leqslant 50$	7
$50 < t \leqslant 60$	2

C D E F G

Probability 1

The probability that an event will happen is a value from 0 to 1.
The probability tells you how likely the event is to happen.
An event that is CERTAIN to happen has a probability of 1.
An event that is IMPOSSIBLE has a probability of 0.
You can write a probability as a fraction, a decimal or a percentage.

Impossible — Even chance — Certain
0 — 1

Fraction	Decimal	Percentage
$\frac{1}{2}$	0.5	50%

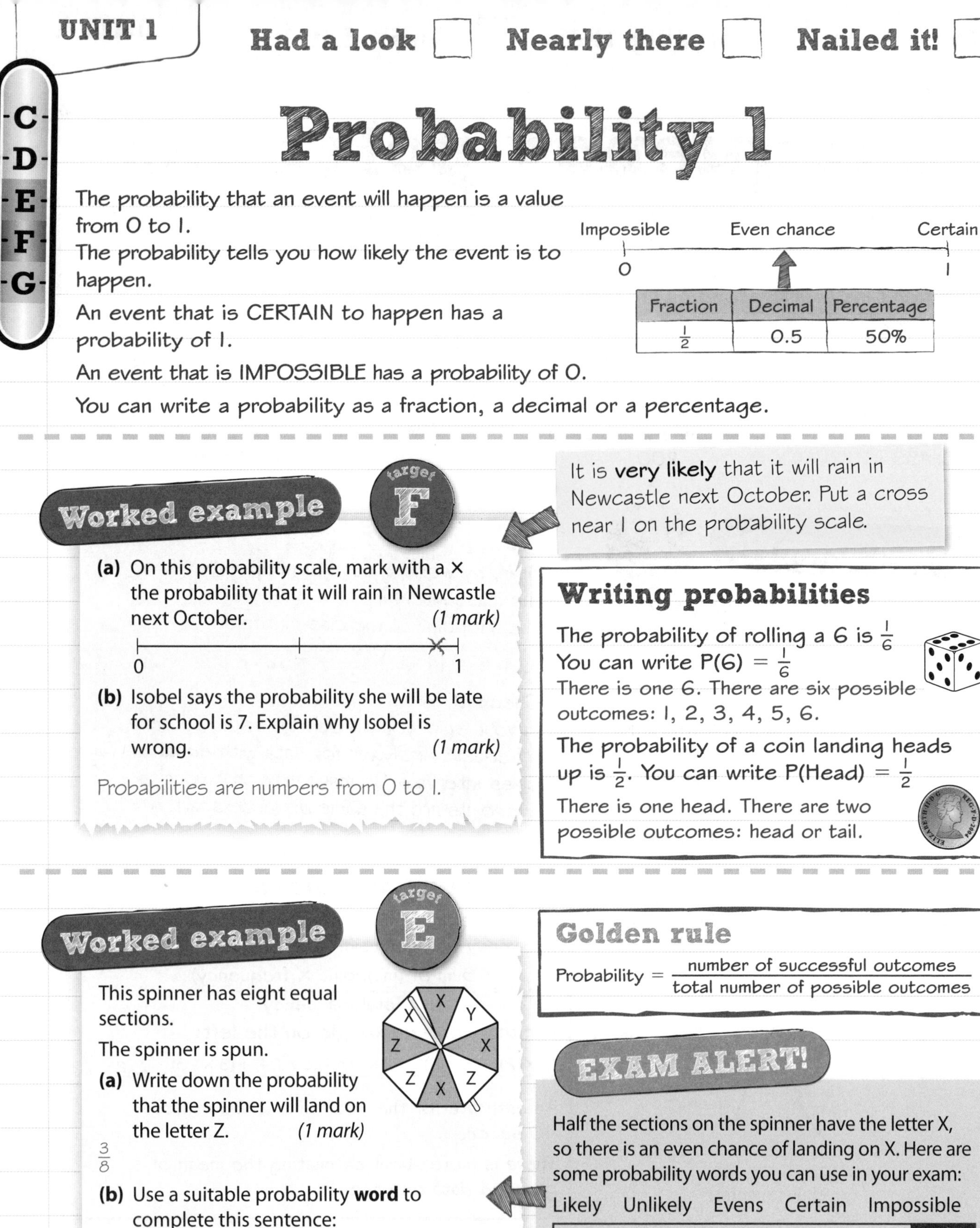

Worked example (target F)

(a) On this probability scale, mark with a × the probability that it will rain in Newcastle next October. *(1 mark)*

0 ——————×— 1

(b) Isobel says the probability she will be late for school is 7. Explain why Isobel is wrong. *(1 mark)*

Probabilities are numbers from 0 to 1.

It is **very likely** that it will rain in Newcastle next October. Put a cross near 1 on the probability scale.

Writing probabilities

The probability of rolling a 6 is $\frac{1}{6}$
You can write $P(6) = \frac{1}{6}$
There is one 6. There are six possible outcomes: 1, 2, 3, 4, 5, 6.

The probability of a coin landing heads up is $\frac{1}{2}$. You can write $P(\text{Head}) = \frac{1}{2}$
There is one head. There are two possible outcomes: head or tail.

Worked example (target E)

This spinner has eight equal sections.
The spinner is spun.

(a) Write down the probability that the spinner will land on the letter Z. *(1 mark)*

$\frac{3}{8}$

(b) Use a suitable probability **word** to complete this sentence:
The chance of the spinner landing on X is evens. *(1 mark)*

Golden rule

$$\text{Probability} = \frac{\text{number of successful outcomes}}{\text{total number of possible outcomes}}$$

EXAM ALERT!

Half the sections on the spinner have the letter X, so there is an even chance of landing on X. Here are some probability words you can use in your exam:
Likely Unlikely Evens Certain Impossible

Students have struggled with exam questions similar to this – **be prepared!**

Now try this (target E)

A spinner has six equal sections.
The spinner is spun.

(a) What is the probability of landing on blue? *(1 mark)*
(b) What is the probability of not landing on red? *(1 mark)*

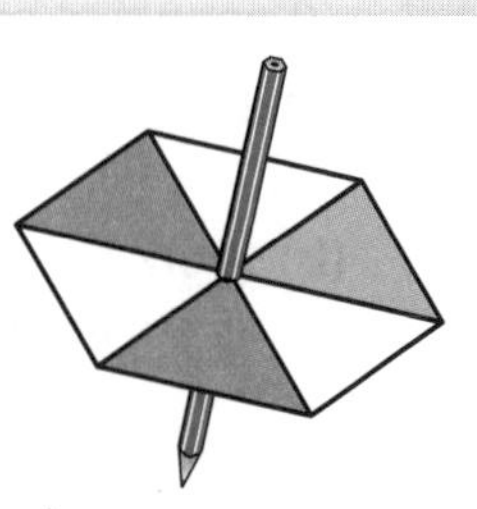

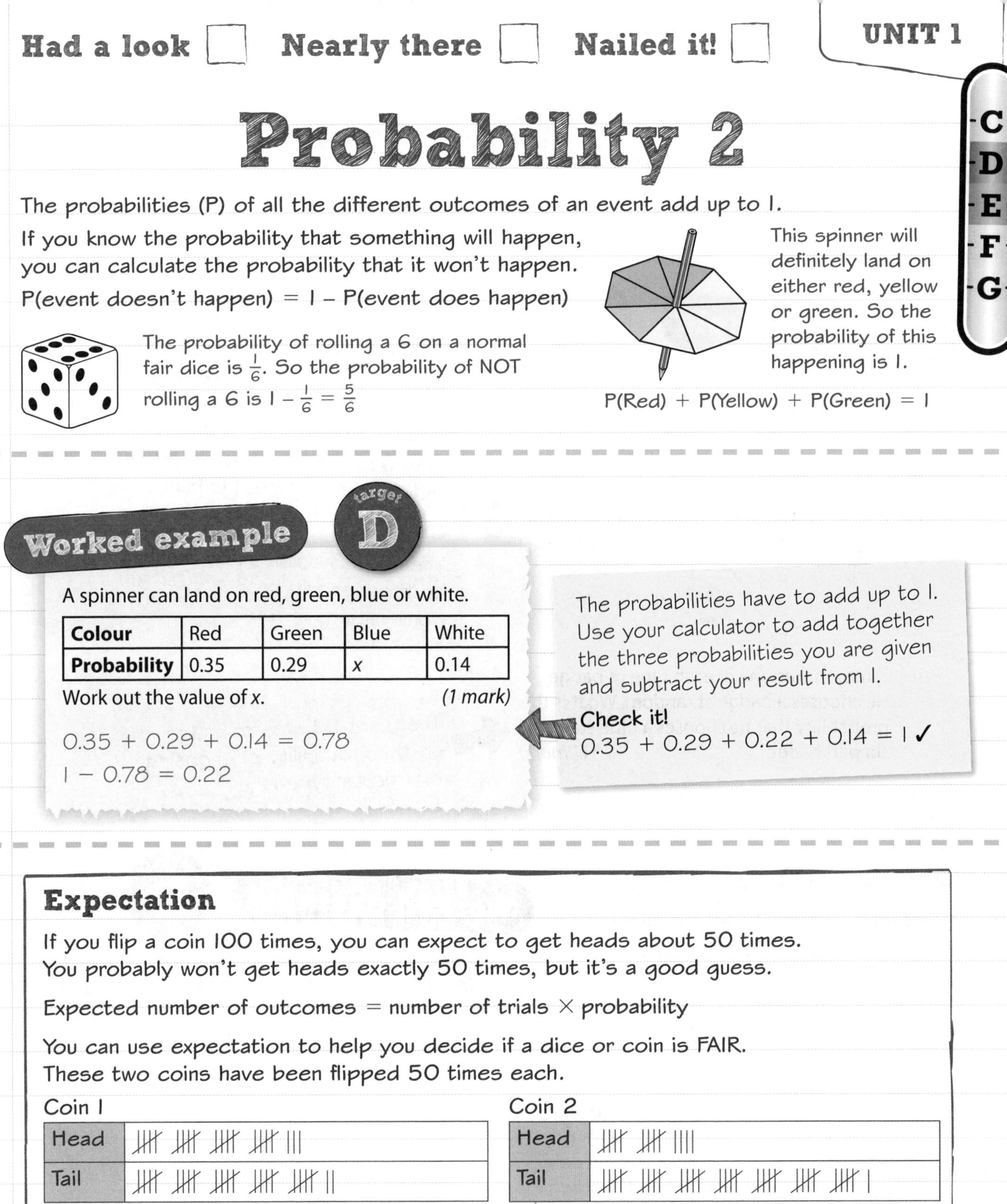

Probability 2

The probabilities (P) of all the different outcomes of an event add up to 1.

If you know the probability that something will happen, you can calculate the probability that it won't happen.

P(event doesn't happen) = 1 – P(event does happen)

The probability of rolling a 6 on a normal fair dice is $\frac{1}{6}$. So the probability of NOT rolling a 6 is $1 - \frac{1}{6} = \frac{5}{6}$

This spinner will definitely land on either red, yellow or green. So the probability of this happening is 1.

P(Red) + P(Yellow) + P(Green) = 1

Worked example

target D

A spinner can land on red, green, blue or white.

Colour	Red	Green	Blue	White
Probability	0.35	0.29	x	0.14

Work out the value of x. *(1 mark)*

0.35 + 0.29 + 0.14 = 0.78

1 – 0.78 = 0.22

The probabilities have to add up to 1. Use your calculator to add together the three probabilities you are given and subtract your result from 1.

Check it!

0.35 + 0.29 + 0.22 + 0.14 = 1 ✓

Expectation

If you flip a coin 100 times, you can expect to get heads about 50 times. You probably won't get heads exactly 50 times, but it's a good guess.

Expected number of outcomes = number of trials × probability

You can use expectation to help you decide if a dice or coin is FAIR. These two coins have been flipped 50 times each.

Coin 1

Head	卌 卌 卌 卌 |||
Tail	卌 卌 卌 卌 卌 ||

About the same number of heads and tails. This coin is probably FAIR.

Coin 2

Head	卌 卌 ||||
Tail	卌 卌 卌 卌 卌 卌 卌	

A lot more than the expected number of tails. This coin is probably BIASED.

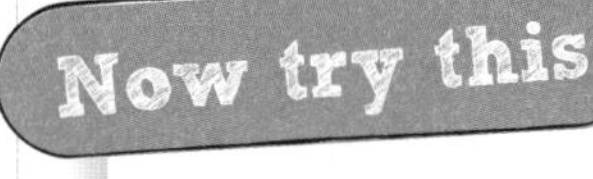

Oscar is playing a game with a **biased** dice.

The probability that he throws a six is 0.3.

(a) What is the probability that he does **not** throw a six? *(1 mark)*

(b) Oscar throws the dice 80 times.
Work out an estimate for the number of sixes he should expect to throw. *(2 marks)*

C D E F G

Combinations

You might need to list all the possible combinations or OUTCOMES of two or more events. You need to find a SYSTEMATIC way of doing this to make sure you have found every possible combination.

target E

James is making badges to sell. He can choose three different colours.

Red (R) Green (G) Blue (B)

He can also choose three shapes.

Circle (C) Star (S) Oval (O)

(a) List **all** the possible combinations of badge colour and shape. *(3 marks)*

RC RS RO
GC GS GO
BC BS BO

(b) James makes one of each type of badge. He chooses a badge at random. What is the probability that he chooses a blue star-shaped badge? *(1 mark)*

$\frac{1}{9}$

You need to use a systematic method to make sure you have written down every possible combination. Start by writing down all the possible **red** badges. Then write down all the possible **green** badges and then all the possible **blue** badges.

Check it!

There are 3 colours and 3 shapes, so there should be $3 \times 3 = 9$ combinations in total.

There are 9 types of badge so the probability of choosing 1 particular type is $\frac{1}{9}$.

Sample space diagrams

A SAMPLE SPACE DIAGRAM shows you all the possible outcomes of two events.

Here are all the possible outcomes when two coins are flipped.

First coin / Second coin

	H	T
H	HH	TH
T	HT	TT

There are four possible outcomes. TH means getting a tail on the first coin and a head on the second coin.

Worked example

target D

This bag contains 30 counters. They are all either black or white.

A counter is chosen at random.

The probability that it is black is $\frac{1}{5}$. How many white counters are in the bag? *(2 marks)*

$P(\text{White}) = 1 - \frac{1}{5} = \frac{4}{5}$

$\frac{4}{5} \times 30 = 24$

Check it!

There are $30 - 24 = 6$ black counters in the bag.

$P(\text{Black}) = \frac{6}{30}$ ✓

Now try this

A bag contains red, blue and green counters.

The probability of picking a red counter is $\frac{1}{8}$.

The probability of picking a blue counter is $\frac{1}{3}$.

(a) What is the probability of picking a green counter? *(2 marks)*

(b) There are 48 counters in the bag altogether. Work out the number of each colour. *(3 marks)*

Multiply each probability by 48.

C D E F G

Relative frequency

You need to be able to calculate probabilities for data given in graphs and tables. You can use this formula to estimate a probability from a frequency table:

$$\text{Probability} = \frac{\text{frequency of outcome}}{\text{total frequency}}$$

When a probability is calculated like this it is sometimes called a RELATIVE FREQUENCY.

Golden rule

Probability estimates based on relative frequency are MORE ACCURATE for larger samples (or for more trials in an experiment).

Worked example

An egg farm weighed a sample of 40 eggs. It recorded the results in a frequency table.

Weight, w (g)	Frequency
$45 \leqslant w < 50$	6
$50 \leqslant w < 55$	9
$55 \leqslant w < 60$	15
$60 \leqslant w < 65$	10

(a) Roselle buys some eggs from the farm and picks one at random. Estimate the probability that the egg weighs 55 g or more. *(2 marks)*

$P(w \geqslant 55) \approx \frac{25}{40} = \frac{5}{8}$

In the sample there were 15 + 10 = 25 eggs which weighed 55 g or more. So an estimate for the probability of picking an egg that weighs 55 g or more is $\frac{25}{40}$ or $\frac{5}{8}$.

(b) Comment on the accuracy of your estimate. *(1 mark)*

40 is a fairly small sample size, so the estimate is not very accurate.

Experimental probability

You can carry out an experiment to estimate the probability of something happening. This table shows the results of throwing a drawing pin 60 times.

Number of trials	10	20	30	40	50	60
Frequency of landing point up	8	11	17	25	30	37

After each 10 trials you can calculate the relative frequency of the drawing pin landing point up, and show the results on a RELATIVE FREQUENCY DIAGRAM.

Now try this

target C

A bag contains a large number of balls. Some are black, some are white and some are yellow.

Liz carried out an experiment to try to work out the probability of picking out a yellow ball.

The table shows her results.

Number of trials	10	20	30	40	50	60
Number of yellow balls	3	7	12	17	23	27
Relative frequency of 'yellow'	0.3					

Relative frequency
0, 0.2, 0.4, 0.6, 0.8, 1
0, 10, 20, 30, 40, 50, 60
Number of trials

(a) Complete the table. *(2 marks)*

(b) Plot the results for relative frequency on the graph. *(2 marks)*

(c) There are 100 balls in the bag. Estimate how many of them are yellow. *(2 marks)*

Had a look ☐ Nearly there ☐ Nailed it! ☐

C D E F G

Comparing data

You can use averages like the MEAN or MEDIAN and measures of spread like the RANGE to compare two sets of data. Follow these steps:

1. Calculate the same average and the range for both data sets.
2. Write a sentence for each statistic COMPARING the values for each data set.
3. Only make a statement if you can back it up with STATISTICAL EVIDENCE.

Calculate then compare

You will often have to compare two sets of data presented in different ways. Make sure you calculate the SAME statistics for both data sets. You will get marks for calculating the statistics correctly AND for comparing the data sets.

Worked example

Melissa and Fran want to compare their long jump distances. They each jump five times. Fran's distances were:

263 cm   194 cm   220 cm   305 cm   280 cm

Melissa's distances have a mean of 292 cm and a range of 185 cm.

Compare the distances for Melissa and Fran.

(4 marks)

$263 + 194 + 220 + 305 + 280 = 1262$

Fran's mean $= 1262 \div 5 = 252.4$ cm

Fran's range $= 305 - 194 = 111$ cm

Melissa jumped further on average because she had a larger mean.

Fran's jumps were more consistent because she had a smaller range.

You are given the mean and the range for Melissa's distances, so you need to calculate these statistics for Fran's distances. There is more about calculating the mean and the range on page 16.

You need to interpret your results in the context of the question. Write one sentence for the mean and one sentence for the range.

Writing conclusions

Here are some examples of good sentences comparing data:

Class A had more consistent exam results because they had a smaller range.

Trees in Park B are shorter on average than trees in Park A (smaller median).

The medians were similar, so on average the apples from both farms were the same weight.

Remember statistics can be the same as well as different.

Now try this

Anna and Carla are in the same Maths class.
Every week they take a 20-mark test.
Here are Anna's first eight test scores.

11   15   9   12   13   16   13   15

Carla's mean score in these tests is 11.
Carla's highest score was 14 and her lowest score was 10.
Compare Anna and Carla's test scores. *(4 marks)*

Start by working out Anna's mean score, and the ranges of both sets of scores.

Problem-solving practice 1

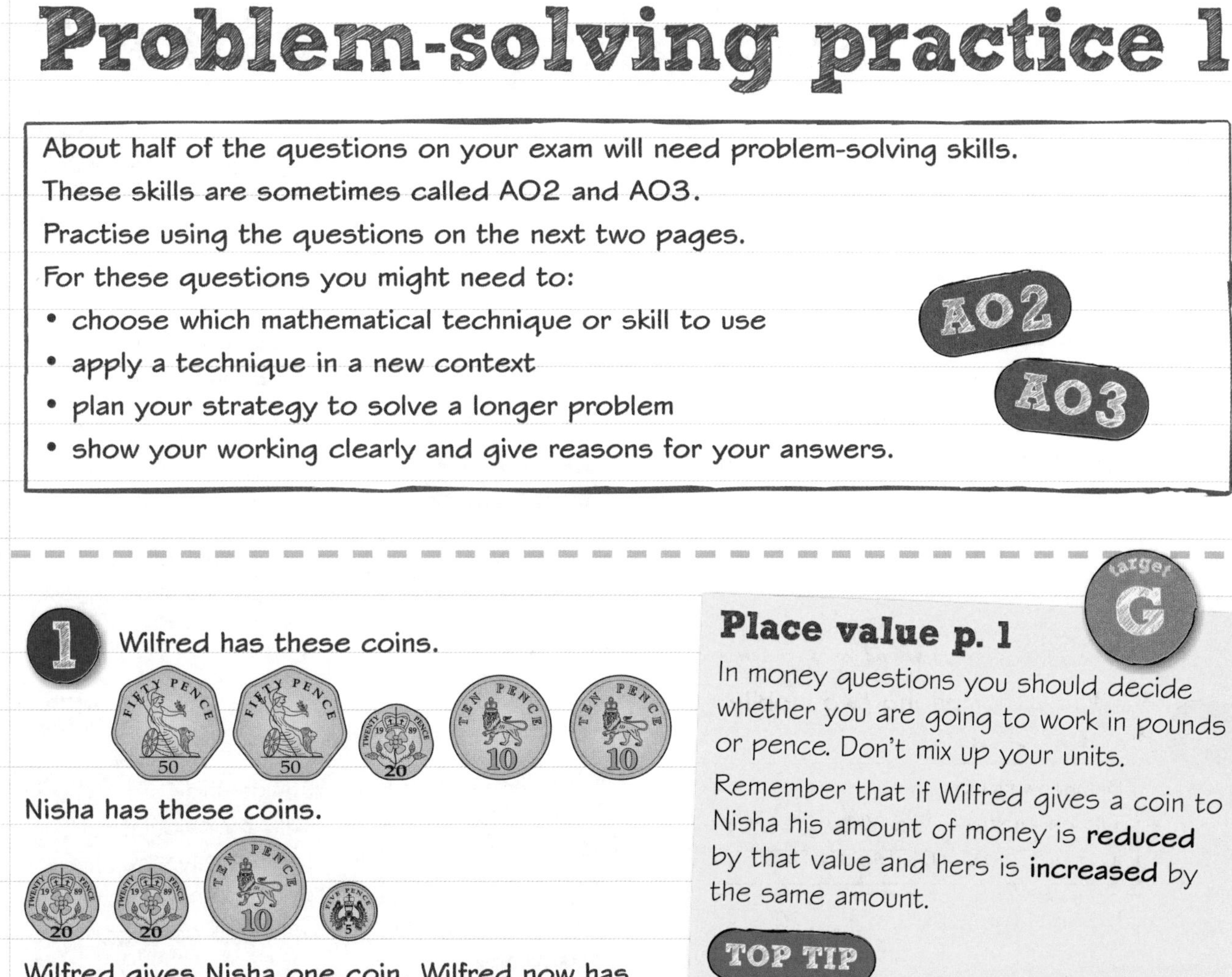

About half of the questions on your exam will need problem-solving skills.
These skills are sometimes called AO2 and AO3.
Practise using the questions on the next two pages.
For these questions you might need to:

- choose which mathematical technique or skill to use
- apply a technique in a new context
- plan your strategy to solve a longer problem
- show your working clearly and give reasons for your answers.

AO2 AO3

1 Wilfred has these coins.

50 50 20 10 10

Nisha has these coins.

20 20 10 5

Wilfred gives Nisha one coin. Wilfred now has twice as much money as Nisha.

What value is the coin Wilfred gives to Nisha?

(2 marks)

target G

Place value p. 1

In money questions you should decide whether you are going to work in pounds or pence. Don't mix up your units.

Remember that if Wilfred gives a coin to Nisha his amount of money is **reduced** by that value and hers is **increased** by the same amount.

TOP TIP

With a question like this it's easy to check your answer. Write down the new amounts of money Wilfred and Nisha have and check that Wilfred's total is twice Nisha's.

2 The table shows information about the numbers of Year 7 pupils absent from Keith's school last week.

	Boys	Girls
Monday	8	10
Tuesday	11	9
Wednesday	12	12
Thursday	14	13
Friday	13	11

Keith wants to compare the data.

Draw a suitable diagram or chart. *(4 marks)*

target F

Bar charts p. 13

In this question **you** have to choose what type of diagram or chart to use. It is best to use a bar chart or a line graph.

TOP TIP

- Label BOTH axes correctly.
- Draw a key for boys and girls, or make sure it is clear which bars (or lines) are for boys and which are for girls.

Problem-solving practice 2

Abi has five cards.

Each card has a number written on it.

The mean of the five numbers is 6

One of the numbers is hidden.

Work out the hidden number. *(2 marks)*

Averages and range p. 16

You could try some different values for the hidden number and work out the mean each time. But you can save time by using the rule in the Top tip. There are five numbers and the mean is 6, so the sum of the numbers must be 5 × 6 = 30.

TOP TIP

This is a useful rule:

mean × number of data values = sum of data values

*Callum is comparing two mobile phone packages.

Business First
£18 per month + 15% VAT
10% DISCOUNT FOR DIRECT DEBIT

Leisure One
One-off yearly payment of £260.
20% DISCOUNT FOR DIRECT DEBIT

Callum says that 'Business First' is cheaper if he pays by Direct Debit. Is Callum correct? Show all your working.

(5 marks)

Percentage change 1 p. 6

You need to plan your answer and show all your working. Start by calculating the cost of each plan if Callum pays by Direct Debit. Then write a sentence comparing the costs and say whether Callum is correct.

TOP TIP

If a question has a * next to it then one mark is awarded for **quality of written communication**. This means you must show all your working and write your answer clearly with the correct units.

Rania measured the reaction times of her class using a computer program. This frequency polygon shows her result.

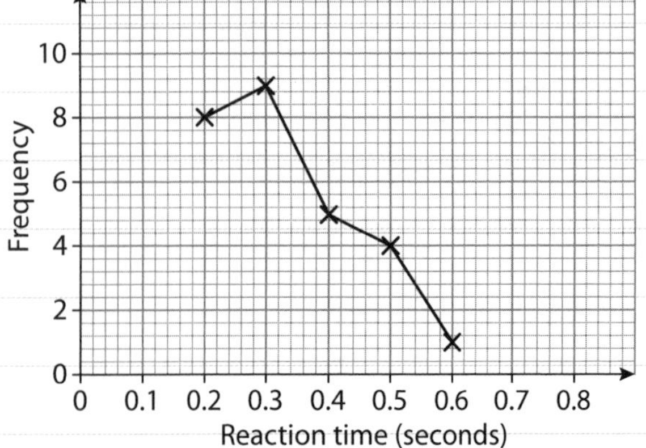

Calculate an estimate of the mean reaction time. *(3 marks)*

Averages from tables 1 p. 18

Frequency polygons are usually used to represent **grouped continuous data**. In a frequency polygon the values are plotted at the **midpoints** of the class intervals, so you have less work to do when estimating the mean.

TOP TIP

Make sure you can read and work with data represented in a table, a graph or as a list of numbers. In your exam you might be given data in any of these formats.

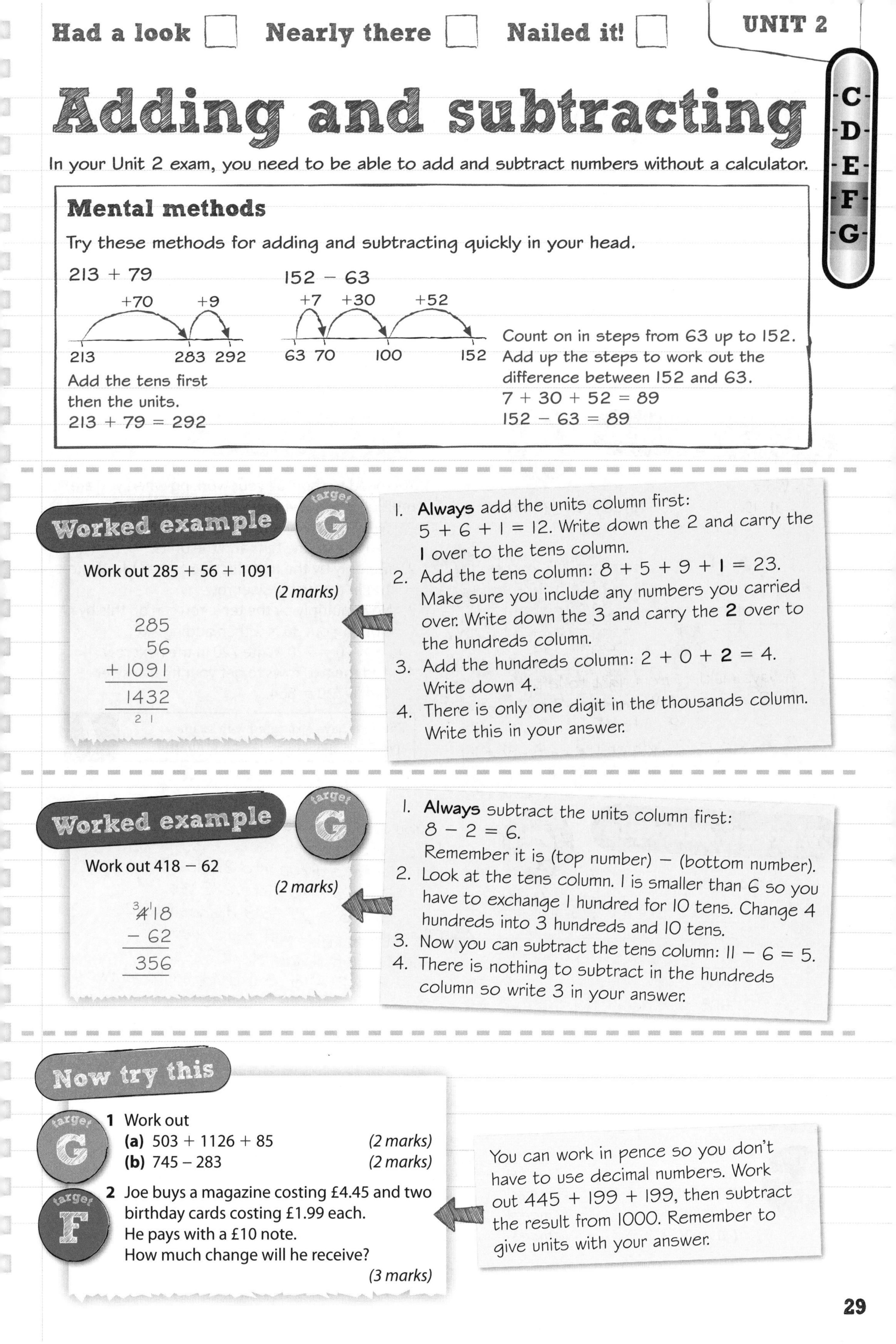

C D E F G

Adding and subtracting

In your Unit 2 exam, you need to be able to add and subtract numbers without a calculator.

Mental methods

Try these methods for adding and subtracting quickly in your head.

213 + 79

+70 +9

213 283 292

Add the tens first then the units.

213 + 79 = 292

152 − 63

+7 +30 +52

63 70 100 152

Count on in steps from 63 up to 152. Add up the steps to work out the difference between 152 and 63.

7 + 30 + 52 = 89

152 − 63 = 89

Worked example

target G

Work out 285 + 56 + 1091

(2 marks)

```
  285
   56
+ 1091
  1432
  2 1
```

1. **Always** add the units column first: 5 + 6 + 1 = 12. Write down the 2 and carry the 1 over to the tens column.
2. Add the tens column: 8 + 5 + 9 + **1** = 23. Make sure you include any numbers you carried over. Write down the 3 and carry the **2** over to the hundreds column.
3. Add the hundreds column: 2 + 0 + **2** = 4. Write down 4.
4. There is only one digit in the thousands column. Write this in your answer.

Worked example

target G

Work out 418 − 62

(2 marks)

```
 3 1
  418
 − 62
  356
```

1. **Always** subtract the units column first: 8 − 2 = 6. Remember it is (top number) − (bottom number).
2. Look at the tens column. 1 is smaller than 6 so you have to exchange 1 hundred for 10 tens. Change 4 hundreds into 3 hundreds and 10 tens.
3. Now you can subtract the tens column: 11 − 6 = 5.
4. There is nothing to subtract in the hundreds column so write 3 in your answer.

Now try this

target G

1 Work out

(a) 503 + 1126 + 85 *(2 marks)*

(b) 745 − 283 *(2 marks)*

target F

2 Joe buys a magazine costing £4.45 and two birthday cards costing £1.99 each.
He pays with a £10 note.
How much change will he receive? *(3 marks)*

You can work in pence so you don't have to use decimal numbers. Work out 445 + 199 + 199, then subtract the result from 1000. Remember to give units with your answer.

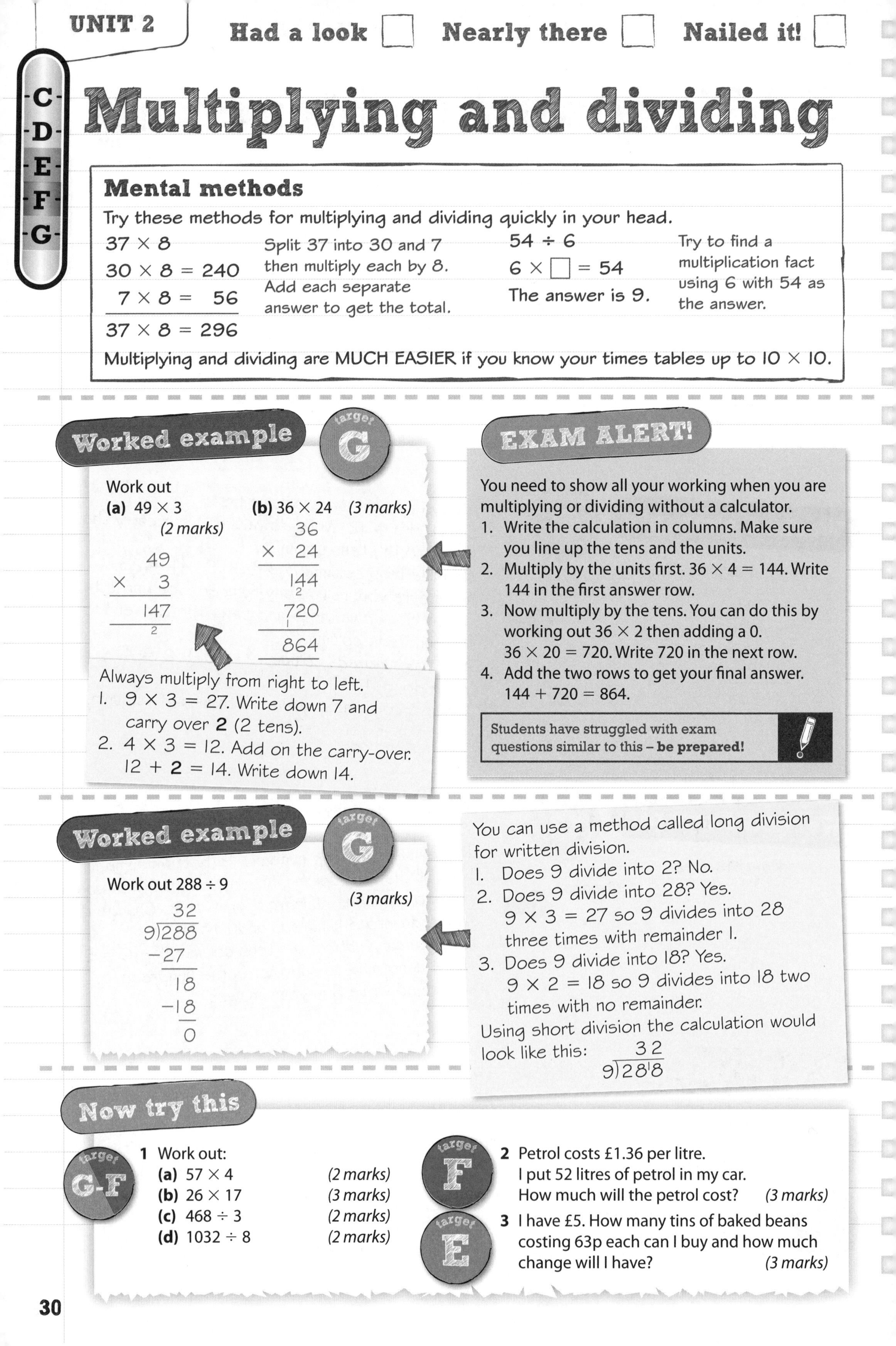

C D E F G

Multiplying and dividing

Mental methods

Try these methods for multiplying and dividing quickly in your head.

37×8
$30 \times 8 = 240$
$7 \times 8 = 56$
$37 \times 8 = 296$

Split 37 into 30 and 7 then multiply each by 8. Add each separate answer to get the total.

$54 \div 6$
$6 \times \square = 54$
The answer is 9.

Try to find a multiplication fact using 6 with 54 as the answer.

Multiplying and dividing are MUCH EASIER if you know your times tables up to 10×10.

Worked example (target G)

Work out
(a) 49×3 *(2 marks)*
(b) 36×24 *(3 marks)*

(a)
```
  49
×  3
 147
  2
```

(b)
```
   36
×  24
  144
   2
  720
   1
  864
```

Always multiply from right to left.
1. $9 \times 3 = 27$. Write down 7 and carry over **2** (2 tens).
2. $4 \times 3 = 12$. Add on the carry-over. $12 + \mathbf{2} = 14$. Write down 14.

EXAM ALERT!

You need to show all your working when you are multiplying or dividing without a calculator.
1. Write the calculation in columns. Make sure you line up the tens and the units.
2. Multiply by the units first. $36 \times 4 = 144$. Write 144 in the first answer row.
3. Now multiply by the tens. You can do this by working out 36×2 then adding a 0. $36 \times 20 = 720$. Write 720 in the next row.
4. Add the two rows to get your final answer. $144 + 720 = 864$.

Students have struggled with exam questions similar to this – **be prepared!**

Worked example (target G)

Work out $288 \div 9$ *(3 marks)*

```
   32
9)288
 -27
   18
  -18
    0
```

You can use a method called long division for written division.
1. Does 9 divide into 2? No.
2. Does 9 divide into 28? Yes. $9 \times 3 = 27$ so 9 divides into 28 three times with remainder 1.
3. Does 9 divide into 18? Yes. $9 \times 2 = 18$ so 9 divides into 18 two times with no remainder.

Using short division the calculation would look like this:

```
   3 2
9)28¹8
```

Now try this

1 Work out: (target G-F)
(a) 57×4 *(2 marks)*
(b) 26×17 *(3 marks)*
(c) $468 \div 3$ *(2 marks)*
(d) $1032 \div 8$ *(2 marks)*

2 Petrol costs £1.36 per litre. I put 52 litres of petrol in my car. How much will the petrol cost? *(3 marks)* (target F)

3 I have £5. How many tins of baked beans costing 63p each can I buy and how much change will I have? *(3 marks)* (target E)

C D E F G

Decimals and place value

You can use a place value diagram to help you understand and compare decimal numbers. Remember that decimal numbers with more digits are not necessarily bigger. Try writing extra 0s so that all the numbers have the same number of decimal places.

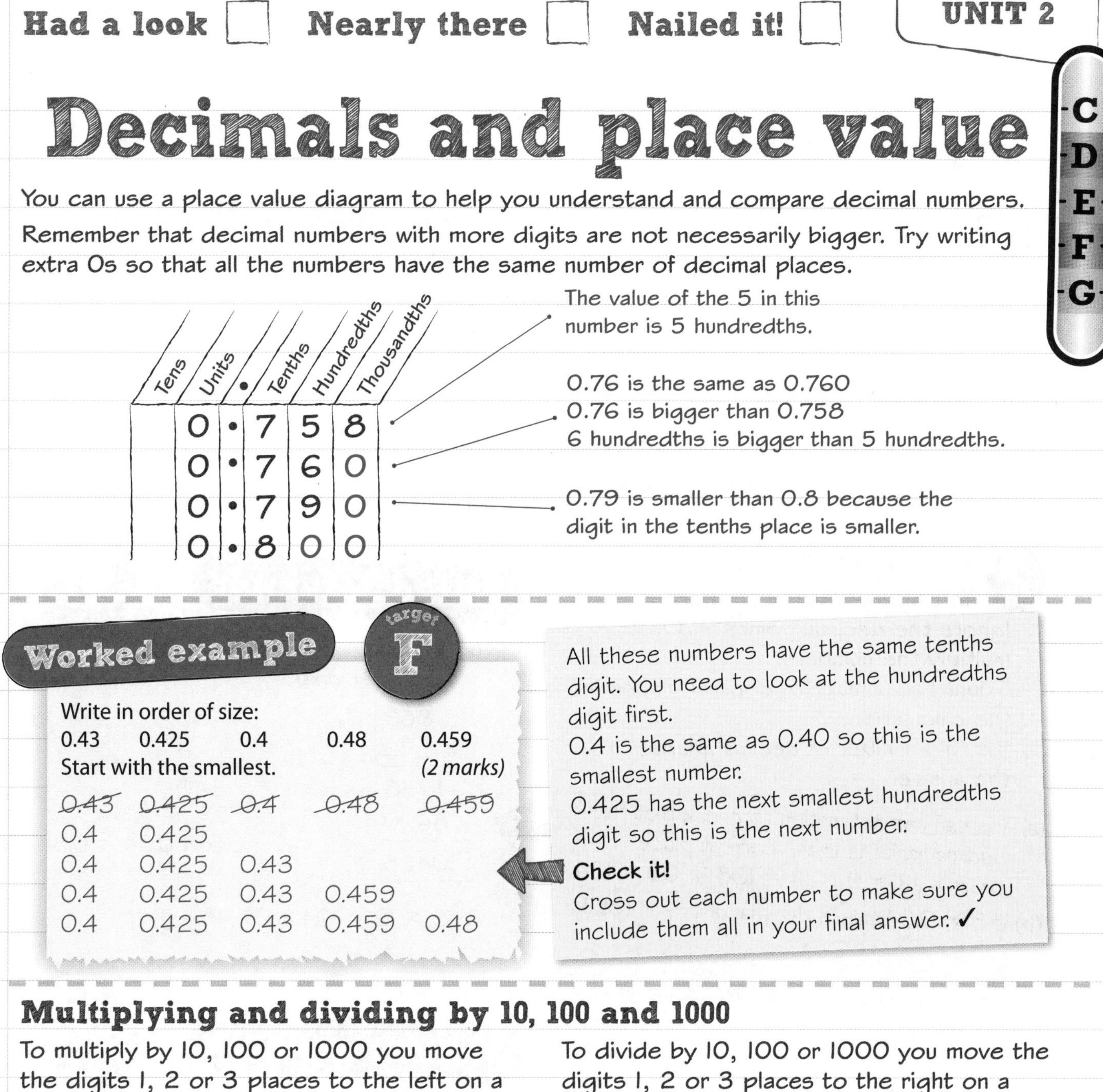

Worked example

target F

Write in order of size:

0.43 0.425 0.4 0.48 0.459

Start with the smallest. *(2 marks)*

~~0.43~~ ~~0.425~~ ~~0.4~~ ~~0.48~~ ~~0.459~~

0.4 0.425

0.4 0.425 0.43

0.4 0.425 0.43 0.459

0.4 0.425 0.43 0.459 0.48

All these numbers have the same tenths digit. You need to look at the hundredths digit first.

0.4 is the same as 0.40 so this is the smallest number.

0.425 has the next smallest hundredths digit so this is the next number.

Check it!

Cross out each number to make sure you include them all in your final answer. ✓

Multiplying and dividing by 10, 100 and 1000

To multiply by 10, 100 or 1000 you move the digits 1, 2 or 3 places to the left on a place value diagram.

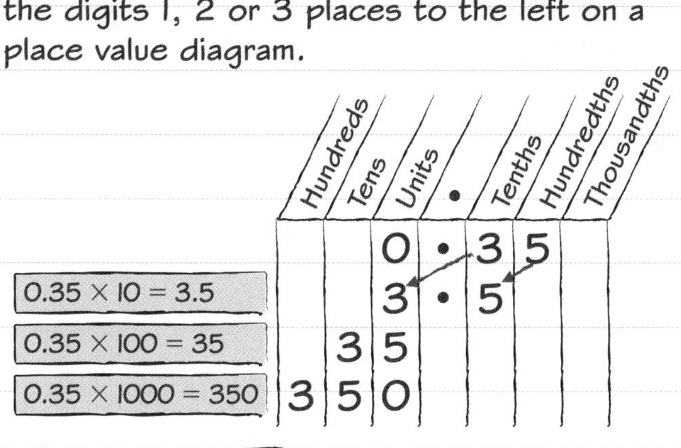

To divide by 10, 100 or 1000 you move the digits 1, 2 or 3 places to the right on a place value diagram.

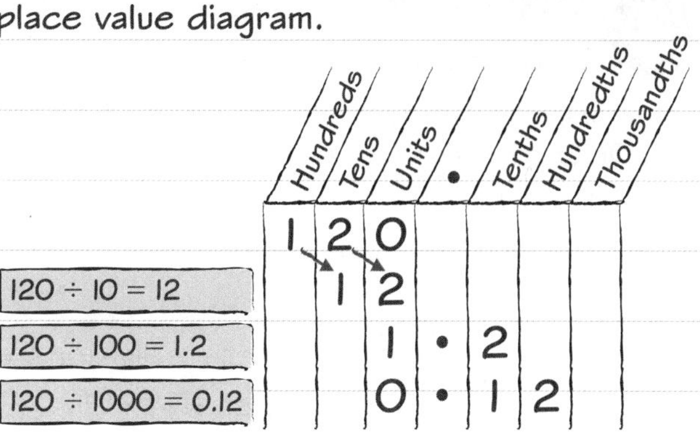

Now try this

target F

1 Write down the place value of the 8 in these numbers.

(a) 2.84 *(1 mark)* **(b)** 0.3086 *(1 mark)*

2 Write these numbers in order of size, starting with the smallest.

0.517 0.508 0.58 0.571 0.51 *(2 marks)*

3 Given that 672 × 13 = 8736, write down the value of

(a) 0.672 × 13 *(1 mark)*

(b) 8736 ÷ 67.2 *(1 mark)*

C D E F G

Operations on decimals

1 Adding and subtracting

To add or subtract decimal numbers:

1. Line up digits with the same place value.
2. Line up the decimal points.
3. Write a decimal point in your answer.

See page 29 for a reminder about adding and subtracting.

Write in 0s so that both numbers have the same number of decimal places.

Worked example (target F-E)

(a) 0.75 + 1.6 *(2 marks)*

```
  0.75
+ 1.60
  2.35
```

(b) 3.5 − 0.21 *(2 marks)*

```
  3.50
- 0.21
  3.29
```

2 Multiplying

To multiply decimal numbers:

1. Ignore the decimal points and just multiply the numbers.
2. Count the number of decimal places in the calculation.
3. Put this number of decimal places in the answer.

(a) You can use estimation to check that the decimal point is in the correct place.
$8.69 \times 12 \approx 9 \times 12 = 108$ $108 \approx 104$ ✓

(b) 8.5×0.04 has 3 decimal places in total

.340 = 0.34

Write a 0 before the decimal point and simplify your answer.

Worked example (target E)

(a) 8.69 × 12 *(3 marks)*

```
   869
×   12
  1738
+ 8690
 10428
```

$8.69 \times 12 = 104.28$

(b) 8.5 × 0.04 *(2 marks)*

```
  85
×  4
 340
```

$8.5 \times 0.04 = 0.34$

3 Dividing

To divide by a decimal number:

1. Multiply both numbers by 10, 100 or 1000 to make the second number a whole number.
2. Divide by the whole number.

Multiply 40.6 and 1.4 by 10.
If you multiply both numbers in a division by the same amount, the answer stays the same.

Worked example (target E-D)

(a) 55.8 ÷ 3 *(2 marks)*

```
   1 8.6
3)5²5.¹8
```

(b) 40.6 ÷ 1.4 *(3 marks)*

$40.6 \div 1.4 = 406 \div 14$

```
     29
14)406
  -28
   126
  -126
     0
```

$40.6 \div 1.4 = 29$

Now try this

Make sure you line up the decimal points.

(target E-D)

1 **(a)** 3.26 + 0.894 + 11.3 *(2 marks)*
(b) 16.5 − 9.72 *(2 marks)*
(c) 5.76 × 34 *(3 marks)*
(d) 456 ÷ 1.2 *(3 marks)*

(target E)

2 **(a)** A kitchen stool costs £39.90.
Work out the cost of six kitchen stools. *(2 marks)*
(b) Tom buys a pack of twelve pencils for £11.28.
Work out the cost of one pencil. *(2 marks)*

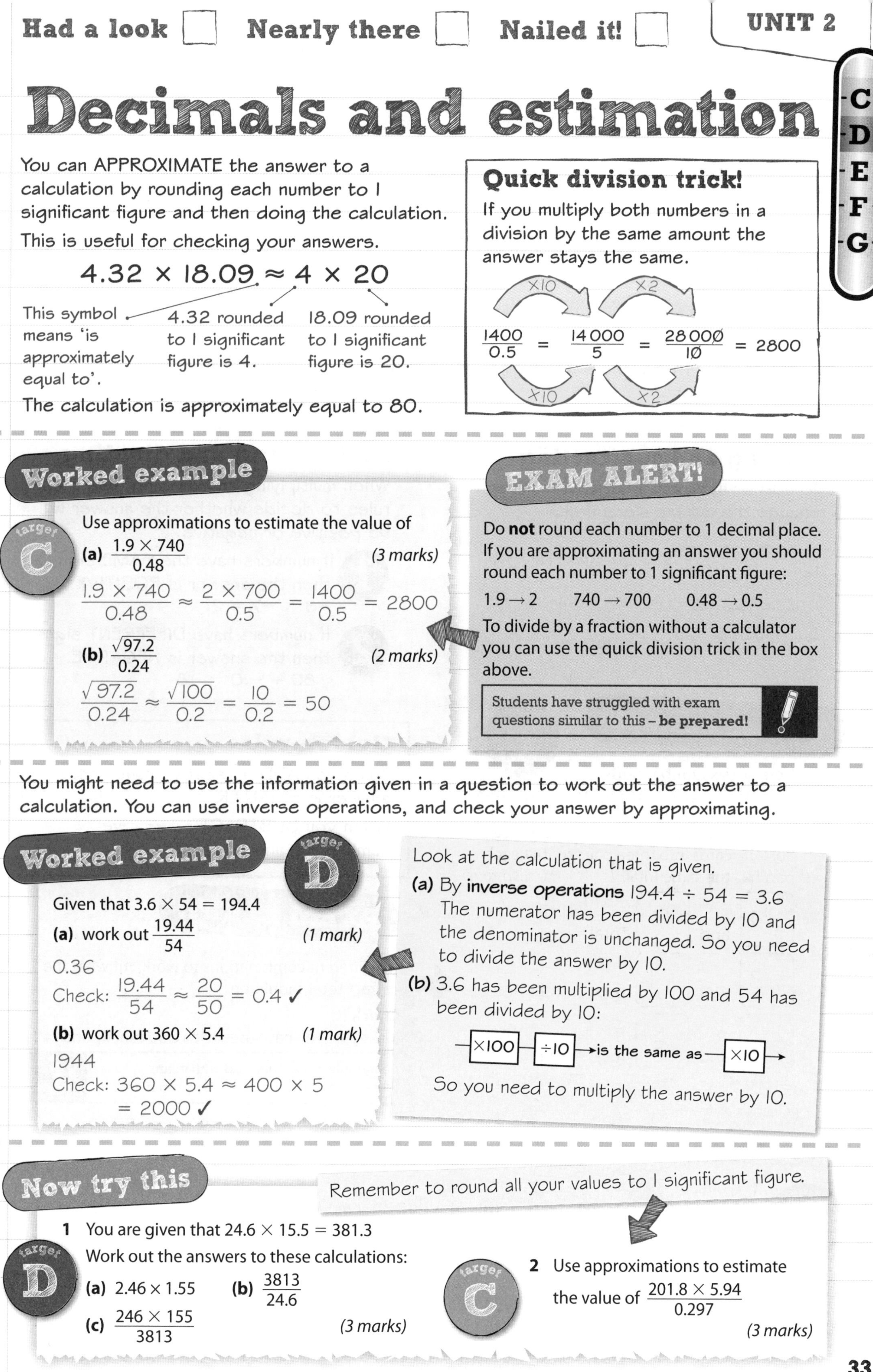

Decimals and estimation

You can APPROXIMATE the answer to a calculation by rounding each number to 1 significant figure and then doing the calculation. This is useful for checking your answers.

$$4.32 \times 18.09 \approx 4 \times 20$$

- This symbol means 'is approximately equal to'.
- 4.32 rounded to 1 significant figure is 4.
- 18.09 rounded to 1 significant figure is 20.

The calculation is approximately equal to 80.

Quick division trick!

If you multiply both numbers in a division by the same amount the answer stays the same.

$$\frac{1400}{0.5} = \frac{14\,000}{5} = \frac{28\,000}{10} = 2800$$

(×10, ×2)

Worked example

Target C

Use approximations to estimate the value of

(a) $\frac{1.9 \times 740}{0.48}$ *(3 marks)*

$$\frac{1.9 \times 740}{0.48} \approx \frac{2 \times 700}{0.5} = \frac{1400}{0.5} = 2800$$

(b) $\frac{\sqrt{97.2}}{0.24}$ *(2 marks)*

$$\frac{\sqrt{97.2}}{0.24} \approx \frac{\sqrt{100}}{0.2} = \frac{10}{0.2} = 50$$

EXAM ALERT!

Do **not** round each number to 1 decimal place. If you are approximating an answer you should round each number to 1 significant figure:

1.9 → 2 740 → 700 0.48 → 0.5

To divide by a fraction without a calculator you can use the quick division trick in the box above.

Students have struggled with exam questions similar to this – **be prepared!**

You might need to use the information given in a question to work out the answer to a calculation. You can use inverse operations, and check your answer by approximating.

Worked example

Target D

Given that $3.6 \times 54 = 194.4$

(a) work out $\frac{19.44}{54}$ *(1 mark)*

0.36

Check: $\frac{19.44}{54} \approx \frac{20}{50} = 0.4$ ✓

(b) work out 360×5.4 *(1 mark)*

1944

Check: $360 \times 5.4 \approx 400 \times 5$
$= 2000$ ✓

Look at the calculation that is given.

(a) By **inverse operations** $194.4 \div 54 = 3.6$
The numerator has been divided by 10 and the denominator is unchanged. So you need to divide the answer by 10.

(b) 3.6 has been multiplied by 100 and 54 has been divided by 10:

→ ×100 → ÷10 → is the same as → ×10 →

So you need to multiply the answer by 10.

Now try this

Remember to round all your values to 1 significant figure.

Target D

1 You are given that $24.6 \times 15.5 = 381.3$
Work out the answers to these calculations:

(a) 2.46×1.55 **(b)** $\frac{3813}{24.6}$

(c) $\frac{246 \times 155}{3813}$ *(3 marks)*

Target C

2 Use approximations to estimate the value of $\frac{201.8 \times 5.94}{0.297}$ *(3 marks)*

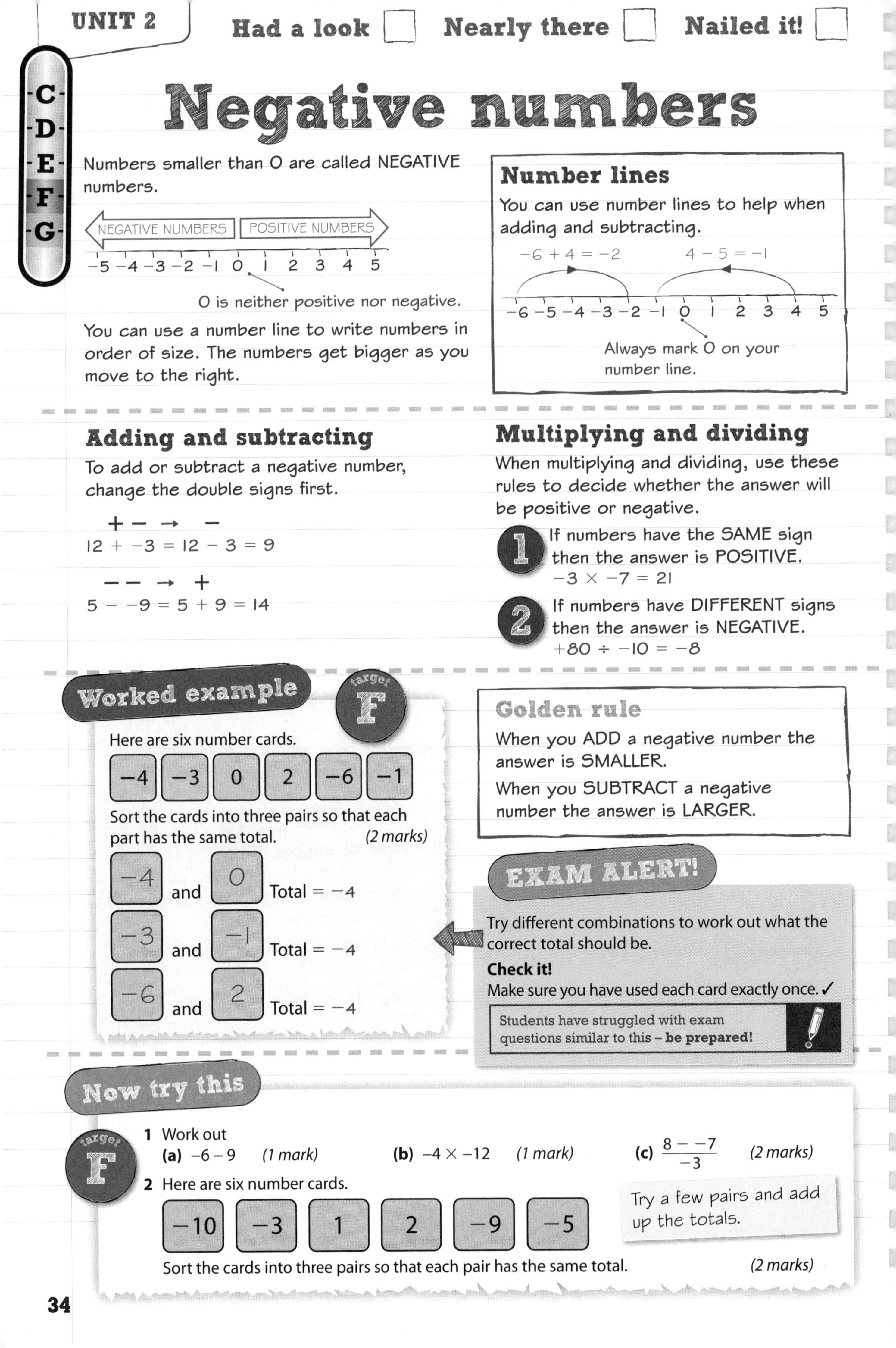

C D E F G

Negative numbers

Numbers smaller than 0 are called NEGATIVE numbers.

You can use a number line to write numbers in order of size. The numbers get bigger as you move to the right.

Number lines

You can use number lines to help when adding and subtracting.

Adding and subtracting

To add or subtract a negative number, change the double signs first.

$+ - \rightarrow -$

$12 + -3 = 12 - 3 = 9$

$- - \rightarrow +$

$5 - -9 = 5 + 9 = 14$

Multiplying and dividing

When multiplying and dividing, use these rules to decide whether the answer will be positive or negative.

1. If numbers have the SAME sign then the answer is POSITIVE.
   $-3 \times -7 = 21$
2. If numbers have DIFFERENT signs then the answer is NEGATIVE.
   $+80 \div -10 = -8$

Worked example

target F

Here are six number cards.

−4	−3	0	2	−6	−1

Sort the cards into three pairs so that each part has the same total. *(2 marks)*

−4 and 0 Total = −4

−3 and −1 Total = −4

−6 and 2 Total = −4

Golden rule

When you ADD a negative number the answer is SMALLER.

When you SUBTRACT a negative number the answer is LARGER.

EXAM ALERT!

Try different combinations to work out what the correct total should be.

Check it!

Make sure you have used each card exactly once. ✓

Students have struggled with exam questions similar to this – **be prepared!**

Now try this

target F

1 Work out

(a) $-6 - 9$ *(1 mark)* **(b)** -4×-12 *(1 mark)* **(c)** $\frac{8 - -7}{-3}$ *(2 marks)*

2 Here are six number cards.

−10	−3	1	2	−9	−5

Sort the cards into three pairs so that each pair has the same total. *(2 marks)*

Try a few pairs and add up the totals.

Squares, cubes and roots

Squares and square roots

When a number is multiplied by itself the answer is a SQUARE NUMBER. You can write square numbers using index notation.

Multiplication	Index notation	Square number
2×2	2^2	4
5×5	5^2	25
9×9	9^2	81
13×13	13^2	169

Square numbers are the areas of squares with whole number side lengths.

25 cm² 5 cm
5 cm

There is more about area on page 80.

SQUARE ROOTS are the opposite of squares. You use the symbol $\sqrt{}$ to represent a square root.

$\sqrt{4} = 2$ $\sqrt{25} = 5$ $\sqrt{81} = 9$

You need to be able to REMEMBER the square numbers up to 15×15 and the corresponding square roots.

Cubes and cube roots

When a number is multiplied by itself then multiplied by itself again, the answer is a CUBE NUMBER. You can write cube numbers using index notation.

Multiplication	Index notation	Cube number
$2 \times 2 \times 2$	2^3	8
$3 \times 3 \times 3$	3^3	27
$5 \times 5 \times 5$	5^3	125
$10 \times 10 \times 10$	10^3	1000

Cube numbers are the volumes of cubes with whole number side lengths.

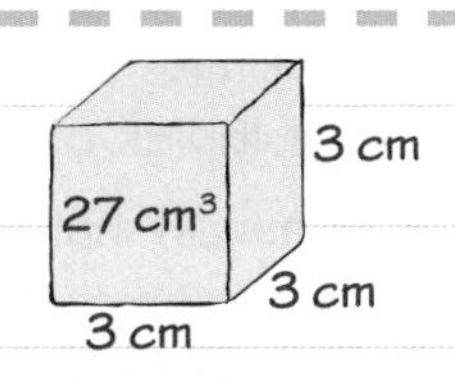

There is more about volume on page 86.

CUBE ROOTS are the opposite of cubes. You use the symbol $\sqrt[3]{}$ to represent a cube root.

$\sqrt[3]{8} = 2$ $\sqrt[3]{27} = 3$ $\sqrt[3]{1000} = 10$

You need to be able to REMEMBER the cubes of 2, 3, 4, 5 and 10, and the corresponding cube roots.

Powers of 10

You can use INDEX NOTATION to write powers of 10. The power tells you how many zeroes the number has.

$10^1 = 10$

$10^2 = 100$

$10^3 = 1000$

$10^4 = 10\,000$

There is more on indices on page 45.

Worked example

target F

Which of these is equal to ten million?
Circle your answer.

10^4 10^5 10^6 (10^7) 10^8 *(1 mark)*

Ten million = 10 000 000

Write down ten million in figures. There are 7 zeroes so the answer is 10^7.

Now try this

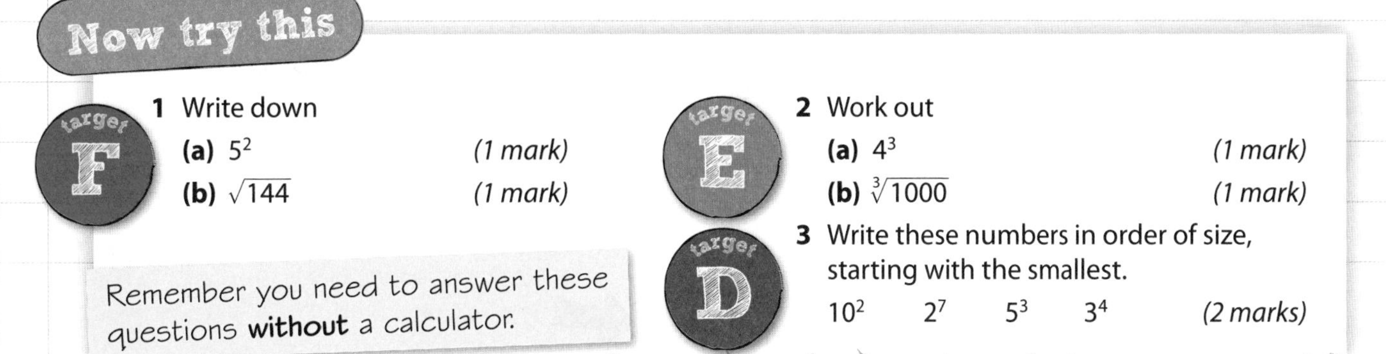

target F

1 Write down

(a) 5^2 *(1 mark)*

(b) $\sqrt{144}$ *(1 mark)*

Remember you need to answer these questions **without** a calculator.

target E

2 Work out

(a) 4^3 *(1 mark)*

(b) $\sqrt[3]{1000}$ *(1 mark)*

target D

3 Write these numbers in order of size, starting with the smallest.

10^2 2^7 5^3 3^4 *(2 marks)*

C D E F G

Factors, multiples and primes

Factors and multiples

The FACTORS of a number are any whole numbers that divide into it exactly.

1 and the number itself are both factors of any number.

The factors of 12 are 1, 2, 3, 4, 6 and 12.

Factors come in pairs. Each pair is a multiplication fact with the number as its answer.

The factor pairs of 12 are 1 × 12, 2 × 6 and 3 × 4.

A common factor is a number that is a factor of two or more numbers.

2 is a common factor of 6 and 12.

The MULTIPLES of a number are all the numbers in its times table.

The multiples of 7 are 7, 14, 21, 28, 35, ...

A common multiple is a number that is a multiple of two or more numbers.

12 is a common multiple of 6 and 4.

Primes

A PRIME NUMBER has exactly two factors. It can only be divided by 1 and by itself.

The first ten prime numbers are 2, 3, 5, 7, 11, 13, 17, 19, 23, 29.

1 is not a prime number. It has only 1 factor.

Factor trees

You can use a factor tree to find prime factors.

1. Choose a factor pair of the number.
2. Circle the prime factors as you go along.
3. Continue until every branch ends with a prime number.
4. At the end write down ALL the circled numbers, putting in multiplication signs.

Worked example (target G-E)

Here is a list of numbers.

16 8 3 17 6 20 12

From this list write down

(a) a prime number *(1 mark)*

17

(b) a multiple of 5 *(1 mark)*

20

(c) two factors of 24 which have a sum of 15. *(2 marks)*

12 and 3

(a) 3 is also a prime number.

(c) 3, 6, 8 and 12 are all factors of 24. Only 12 and 3 have a sum of 15.

Worked example

Write 90 as the product of prime factors. Give your answer in index form. *(3 marks)*

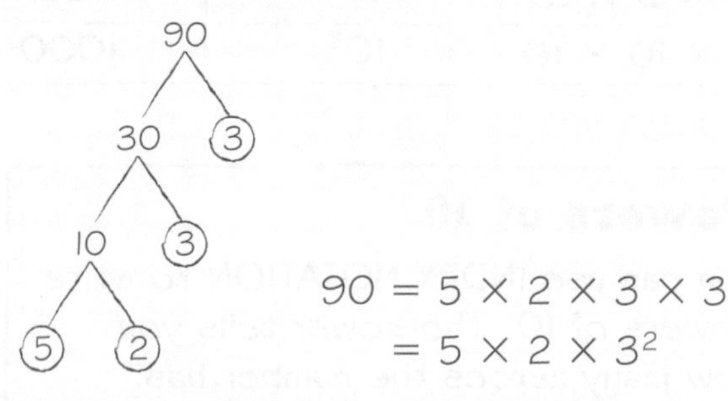

Use a factor tree to find all the prime factors. 3 appears twice in the factor tree, so you have to write 3^2.

Check it!

$5 \times 2 \times 3^2 = 10 \times 9 = 90$ ✓

Now try this

1 Here is a list of numbers.

2 8 15 18 21 24 37 44

From the list, write down

(a) the number that is a multiple of 7 *(1 mark)*

(b) the number that is a factor of 45 *(1 mark)*

(c) the number that is a multiple of 6 and a factor of 48 *(1 mark)*

(d) two prime numbers. *(2 marks)*

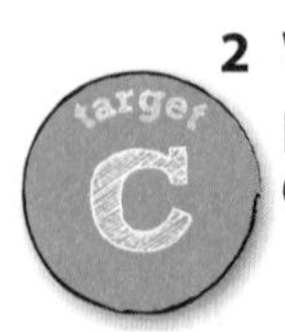

2 Write 280 as the product of its prime factors. Give your answer in index form. *(3 marks)*

Make sure you can recognise all the prime numbers below 50.

Had a look ☐ Nearly there ☐ Nailed it! ☐

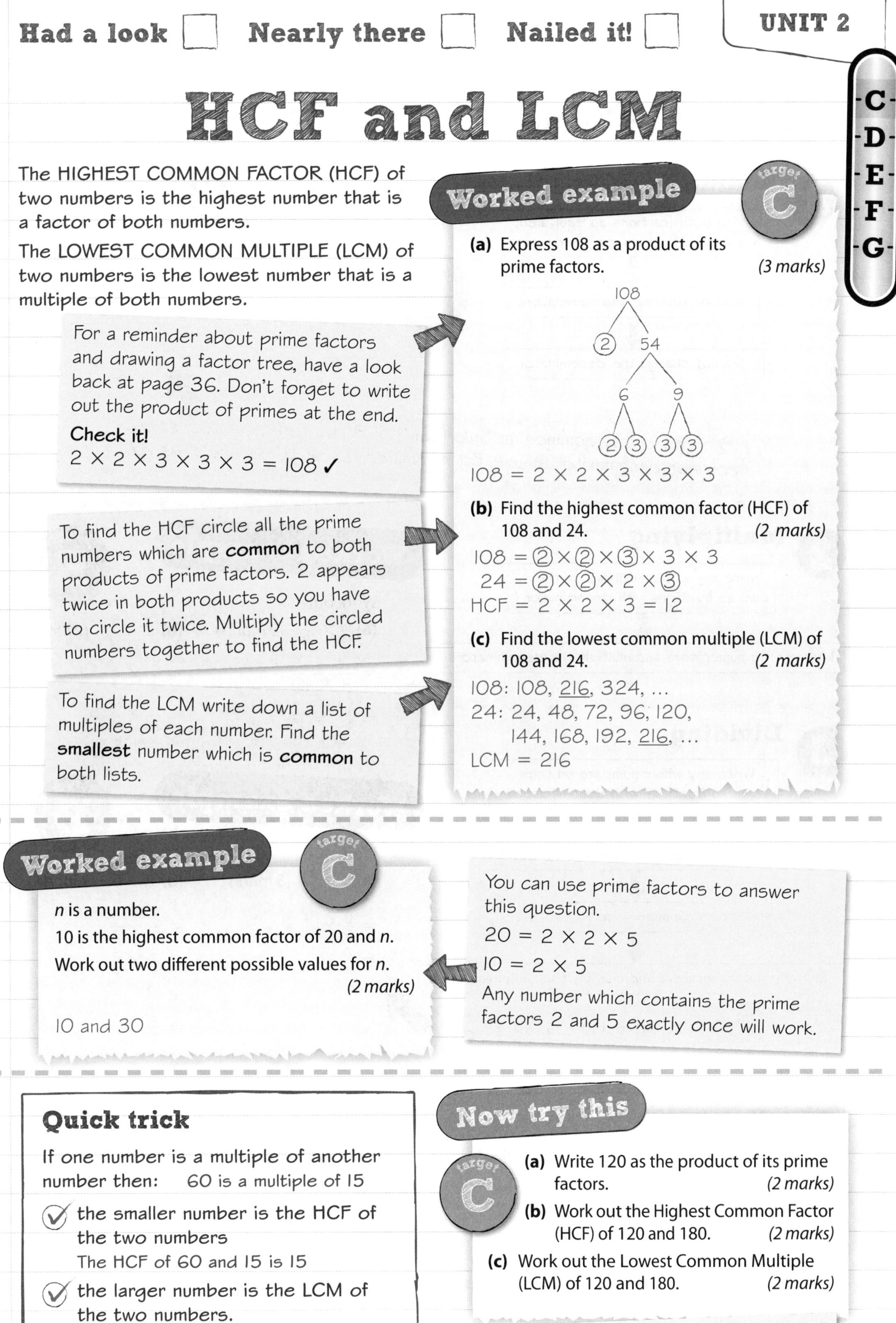

HCF and LCM

The HIGHEST COMMON FACTOR (HCF) of two numbers is the highest number that is a factor of both numbers.

The LOWEST COMMON MULTIPLE (LCM) of two numbers is the lowest number that is a multiple of both numbers.

Worked example

target C

(a) Express 108 as a product of its prime factors. *(3 marks)*

$108 = 2 \times 2 \times 3 \times 3 \times 3$

For a reminder about prime factors and drawing a factor tree, have a look back at page 36. Don't forget to write out the product of primes at the end.

Check it!

$2 \times 2 \times 3 \times 3 \times 3 = 108$ ✓

(b) Find the highest common factor (HCF) of 108 and 24. *(2 marks)*

$108 = (2) \times (2) \times (3) \times 3 \times 3$

$24 = (2) \times (2) \times 2 \times (3)$

$\text{HCF} = 2 \times 2 \times 3 = 12$

To find the HCF circle all the prime numbers which are **common** to both products of prime factors. 2 appears twice in both products so you have to circle it twice. Multiply the circled numbers together to find the HCF.

(c) Find the lowest common multiple (LCM) of 108 and 24. *(2 marks)*

108: 108, 216, 324, …

24: 24, 48, 72, 96, 120, 144, 168, 192, 216, …

LCM = 216

To find the LCM write down a list of multiples of each number. Find the **smallest** number which is **common** to both lists.

Worked example

target C

n is a number.

10 is the highest common factor of 20 and *n*.

Work out two different possible values for *n*. *(2 marks)*

10 and 30

You can use prime factors to answer this question.

$20 = 2 \times 2 \times 5$

$10 = 2 \times 5$

Any number which contains the prime factors 2 and 5 exactly once will work.

Quick trick

If one number is a multiple of another number then: 60 is a multiple of 15

- ✓ the smaller number is the HCF of the two numbers
  The HCF of 60 and 15 is 15
- ✓ the larger number is the LCM of the two numbers.
  The LCM of 60 and 15 is 60

Now try this

target C

(a) Write 120 as the product of its prime factors. *(2 marks)*

(b) Work out the Highest Common Factor (HCF) of 120 and 180. *(2 marks)*

(c) Work out the Lowest Common Multiple (LCM) of 120 and 180. *(2 marks)*

Use a factor tree. Remember to write your final answer using × signs.

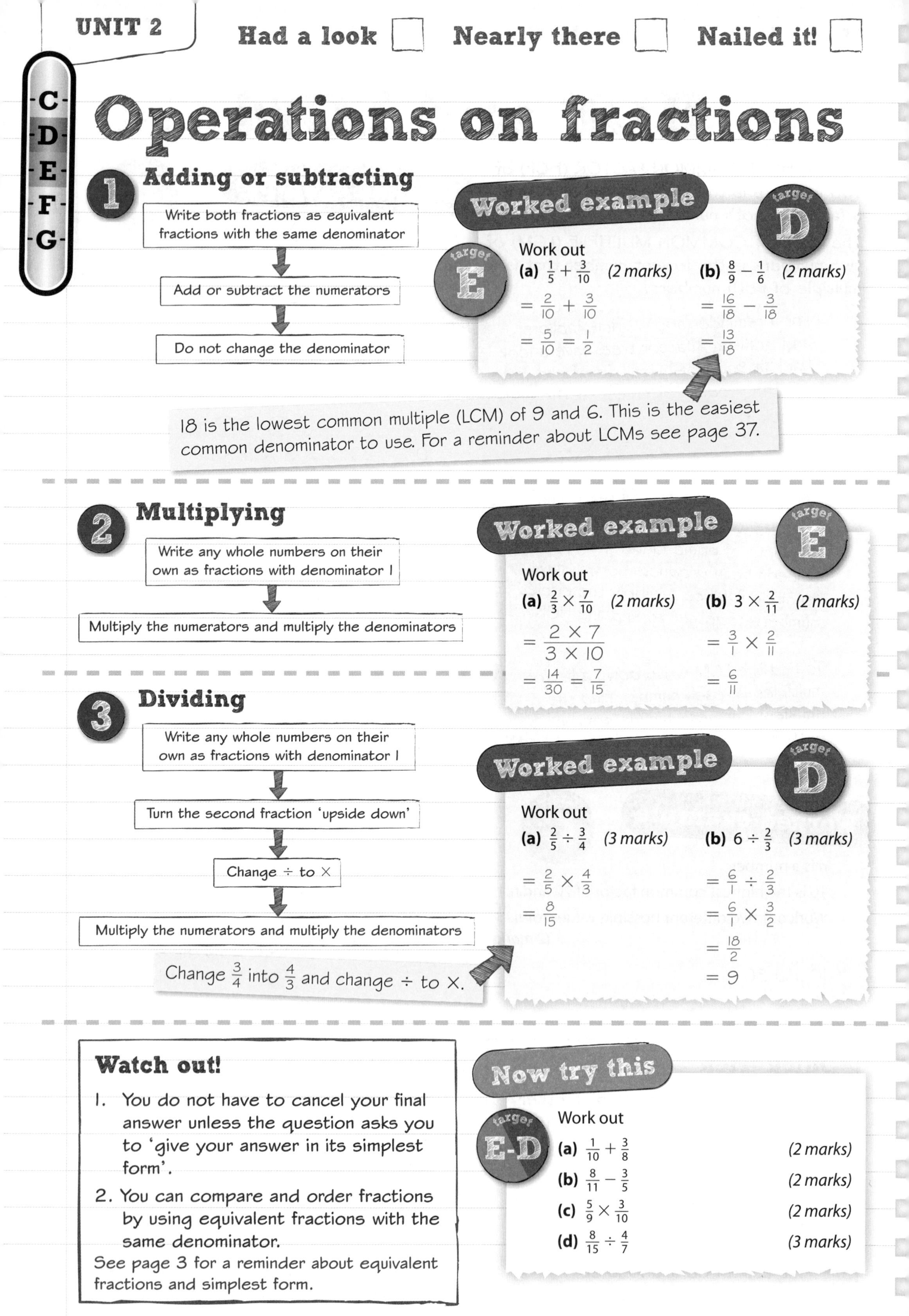

Operations on fractions

C D E F G

1 Adding or subtracting

Write both fractions as equivalent fractions with the same denominator

↓

Add or subtract the numerators

↓

Do not change the denominator

Worked example target E / target D

Work out

(a) $\frac{1}{5} + \frac{3}{10}$ *(2 marks)*

$= \frac{2}{10} + \frac{3}{10}$

$= \frac{5}{10} = \frac{1}{2}$

(b) $\frac{8}{9} - \frac{1}{6}$ *(2 marks)*

$= \frac{16}{18} - \frac{3}{18}$

$= \frac{13}{18}$

18 is the lowest common multiple (LCM) of 9 and 6. This is the easiest common denominator to use. For a reminder about LCMs see page 37.

2 Multiplying

Write any whole numbers on their own as fractions with denominator 1

↓

Multiply the numerators and multiply the denominators

Worked example target E

Work out

(a) $\frac{2}{3} \times \frac{7}{10}$ *(2 marks)*

$= \frac{2 \times 7}{3 \times 10}$

$= \frac{14}{30} = \frac{7}{15}$

(b) $3 \times \frac{2}{11}$ *(2 marks)*

$= \frac{3}{1} \times \frac{2}{11}$

$= \frac{6}{11}$

3 Dividing

Write any whole numbers on their own as fractions with denominator 1

↓

Turn the second fraction 'upside down'

↓

Change ÷ to ×

↓

Multiply the numerators and multiply the denominators

Change $\frac{3}{4}$ into $\frac{4}{3}$ and change ÷ to ×.

Worked example target D

Work out

(a) $\frac{2}{5} \div \frac{3}{4}$ *(3 marks)*

$= \frac{2}{5} \times \frac{4}{3}$

$= \frac{8}{15}$

(b) $6 \div \frac{2}{3}$ *(3 marks)*

$= \frac{6}{1} \div \frac{2}{3}$

$= \frac{6}{1} \times \frac{3}{2}$

$= \frac{18}{2}$

$= 9$

Watch out!

1. You do not have to cancel your final answer unless the question asks you to 'give your answer in its simplest form'.
2. You can compare and order fractions by using equivalent fractions with the same denominator.

See page 3 for a reminder about equivalent fractions and simplest form.

Now try this

target E-D

Work out

(a) $\frac{1}{10} + \frac{3}{8}$ *(2 marks)*

(b) $\frac{8}{11} - \frac{3}{5}$ *(2 marks)*

(c) $\frac{5}{9} \times \frac{3}{10}$ *(2 marks)*

(d) $\frac{8}{15} \div \frac{4}{7}$ *(3 marks)*

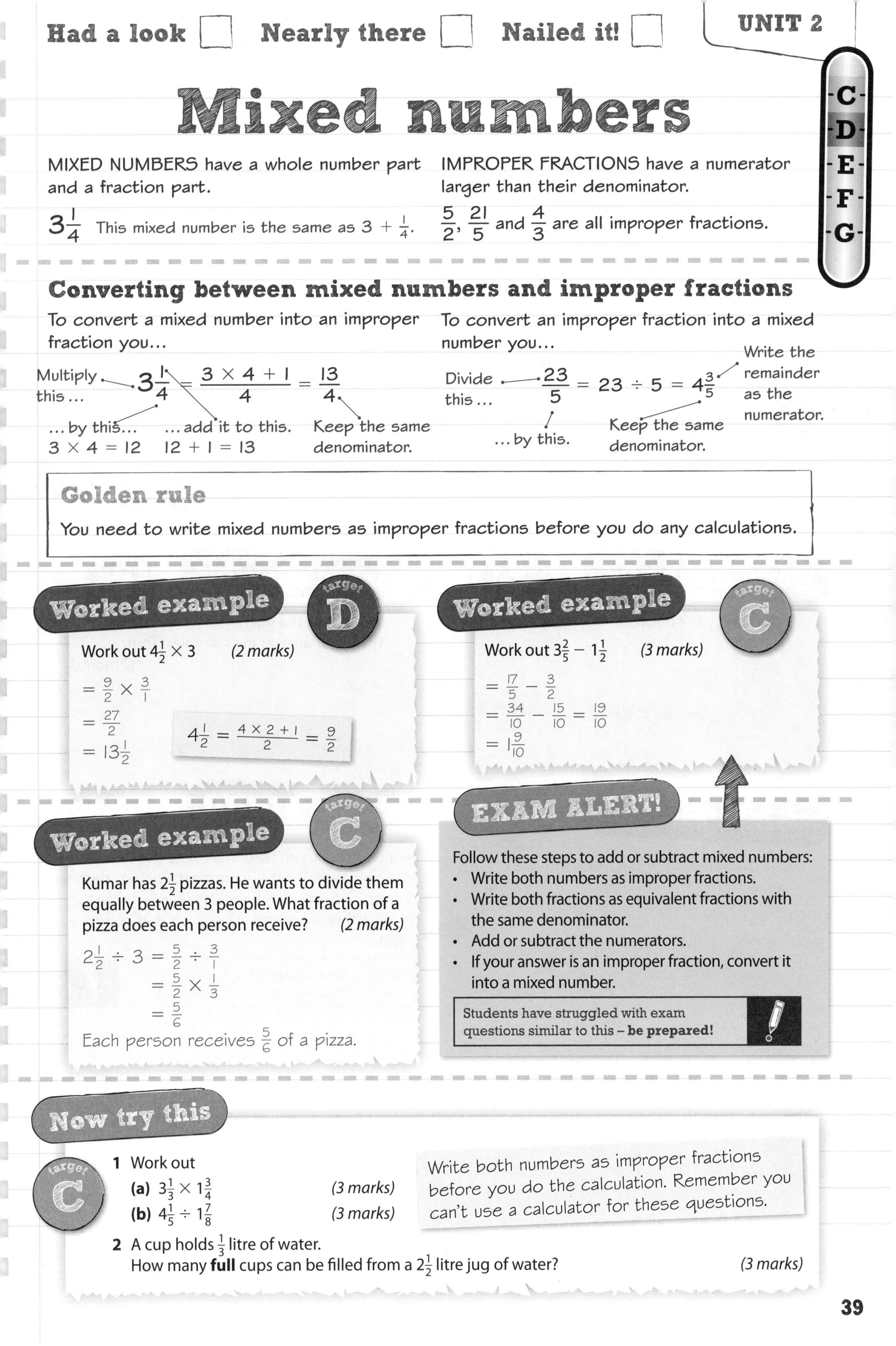

C D E F G

Mixed numbers

MIXED NUMBERS have a whole number part and a fraction part.

$3\frac{1}{4}$ This mixed number is the same as $3 + \frac{1}{4}$.

IMPROPER FRACTIONS have a numerator larger than their denominator.

$\frac{5}{2}$, $\frac{21}{5}$ and $\frac{4}{3}$ are all improper fractions.

Converting between mixed numbers and improper fractions

To convert a mixed number into an improper fraction you…

$$3\frac{1}{4} = \frac{3 \times 4 + 1}{4} = \frac{13}{4}$$

Multiply this … ($3\frac{1}{4}$) … by this… $3 \times 4 = 12$ … add it to this. $12 + 1 = 13$ Keep the same denominator.

To convert an improper fraction into a mixed number you…

$$\frac{23}{5} = 23 \div 5 = 4\frac{3}{5}$$

Divide this … … by this. Keep the same denominator. Write the remainder as the numerator.

Golden rule

You need to write mixed numbers as improper fractions before you do any calculations.

Worked example (target D)

Work out $4\frac{1}{2} \times 3$ *(2 marks)*

$= \frac{9}{2} \times \frac{3}{1}$

$= \frac{27}{2}$

$= 13\frac{1}{2}$

$4\frac{1}{2} = \frac{4 \times 2 + 1}{2} = \frac{9}{2}$

Worked example (target C)

Work out $3\frac{2}{5} - 1\frac{1}{2}$ *(3 marks)*

$= \frac{17}{5} - \frac{3}{2}$

$= \frac{34}{10} - \frac{15}{10} = \frac{19}{10}$

$= 1\frac{9}{10}$

Worked example (target C)

Kumar has $2\frac{1}{2}$ pizzas. He wants to divide them equally between 3 people. What fraction of a pizza does each person receive? *(2 marks)*

$2\frac{1}{2} \div 3 = \frac{5}{2} \div \frac{3}{1}$

$= \frac{5}{2} \times \frac{1}{3}$

$= \frac{5}{6}$

Each person receives $\frac{5}{6}$ of a pizza.

EXAM ALERT!

Follow these steps to add or subtract mixed numbers:

- Write both numbers as improper fractions.
- Write both fractions as equivalent fractions with the same denominator.
- Add or subtract the numerators.
- If your answer is an improper fraction, convert it into a mixed number.

Students have struggled with exam questions similar to this – **be prepared!**

Now try this (target C)

1 Work out

(a) $3\frac{1}{3} \times 1\frac{3}{4}$ *(3 marks)*

(b) $4\frac{1}{5} \div 1\frac{7}{8}$ *(3 marks)*

Write both numbers as improper fractions before you do the calculation. Remember you can't use a calculator for these questions.

2 A cup holds $\frac{1}{3}$ litre of water.
How many **full** cups can be filled from a $2\frac{1}{2}$ litre jug of water? *(3 marks)*

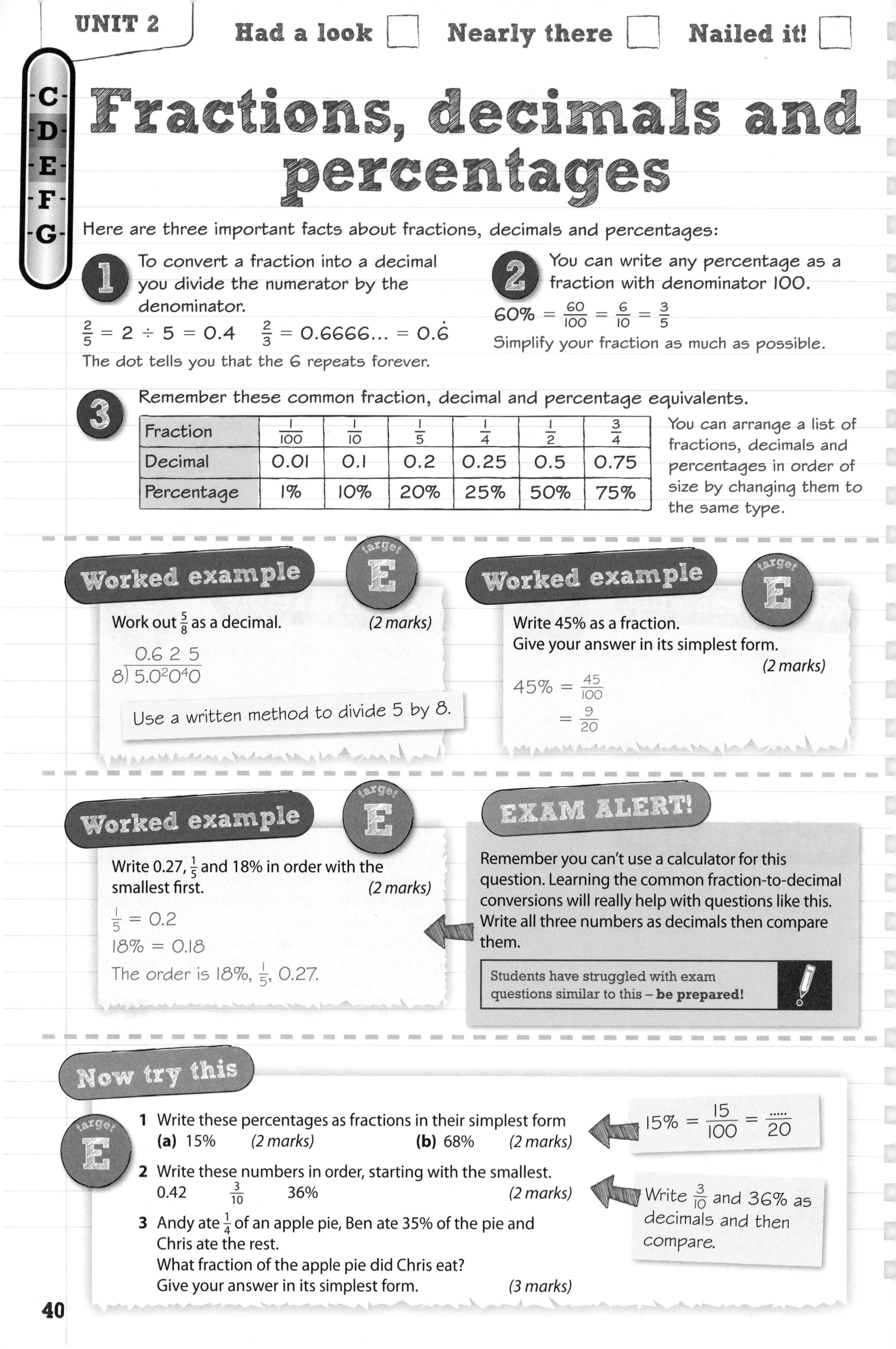

C D E F G

Fractions, decimals and percentages

Here are three important facts about fractions, decimals and percentages:

1 To convert a fraction into a decimal you divide the numerator by the denominator.

$\frac{2}{5} = 2 \div 5 = 0.4$ $\frac{2}{3} = 0.6666... = 0.\dot{6}$

The dot tells you that the 6 repeats forever.

2 You can write any percentage as a fraction with denominator 100.

$60\% = \frac{60}{100} = \frac{6}{10} = \frac{3}{5}$

Simplify your fraction as much as possible.

3 Remember these common fraction, decimal and percentage equivalents.

Fraction	$\frac{1}{100}$	$\frac{1}{10}$	$\frac{1}{5}$	$\frac{1}{4}$	$\frac{1}{2}$	$\frac{3}{4}$
Decimal	0.01	0.1	0.2	0.25	0.5	0.75
Percentage	1%	10%	20%	25%	50%	75%

You can arrange a list of fractions, decimals and percentages in order of size by changing them to the same type.

Worked example (target E)

Work out $\frac{5}{8}$ as a decimal. *(2 marks)*

$$8\overline{)5.0^{2}0^{4}0}$$ = 0.6 2 5

Use a written method to divide 5 by 8.

Worked example (target E)

Write 45% as a fraction.
Give your answer in its simplest form. *(2 marks)*

$45\% = \frac{45}{100}$

$= \frac{9}{20}$

Worked example (target E)

Write 0.27, $\frac{1}{5}$ and 18% in order with the smallest first. *(2 marks)*

$\frac{1}{5} = 0.2$

$18\% = 0.18$

The order is 18%, $\frac{1}{5}$, 0.27.

EXAM ALERT!

Remember you can't use a calculator for this question. Learning the common fraction-to-decimal conversions will really help with questions like this. Write all three numbers as decimals then compare them.

Students have struggled with exam questions similar to this – **be prepared!**

Now try this (target E)

1 Write these percentages as fractions in their simplest form
(a) 15% *(2 marks)* **(b)** 68% *(2 marks)*

$15\% = \frac{15}{100} = \frac{.....}{20}$

2 Write these numbers in order, starting with the smallest.
0.42 $\frac{3}{10}$ 36% *(2 marks)*

Write $\frac{3}{10}$ and 36% as decimals and then compare.

3 Andy ate $\frac{1}{4}$ of an apple pie, Ben ate 35% of the pie and Chris ate the rest.
What fraction of the apple pie did Chris eat?
Give your answer in its simplest form. *(3 marks)*

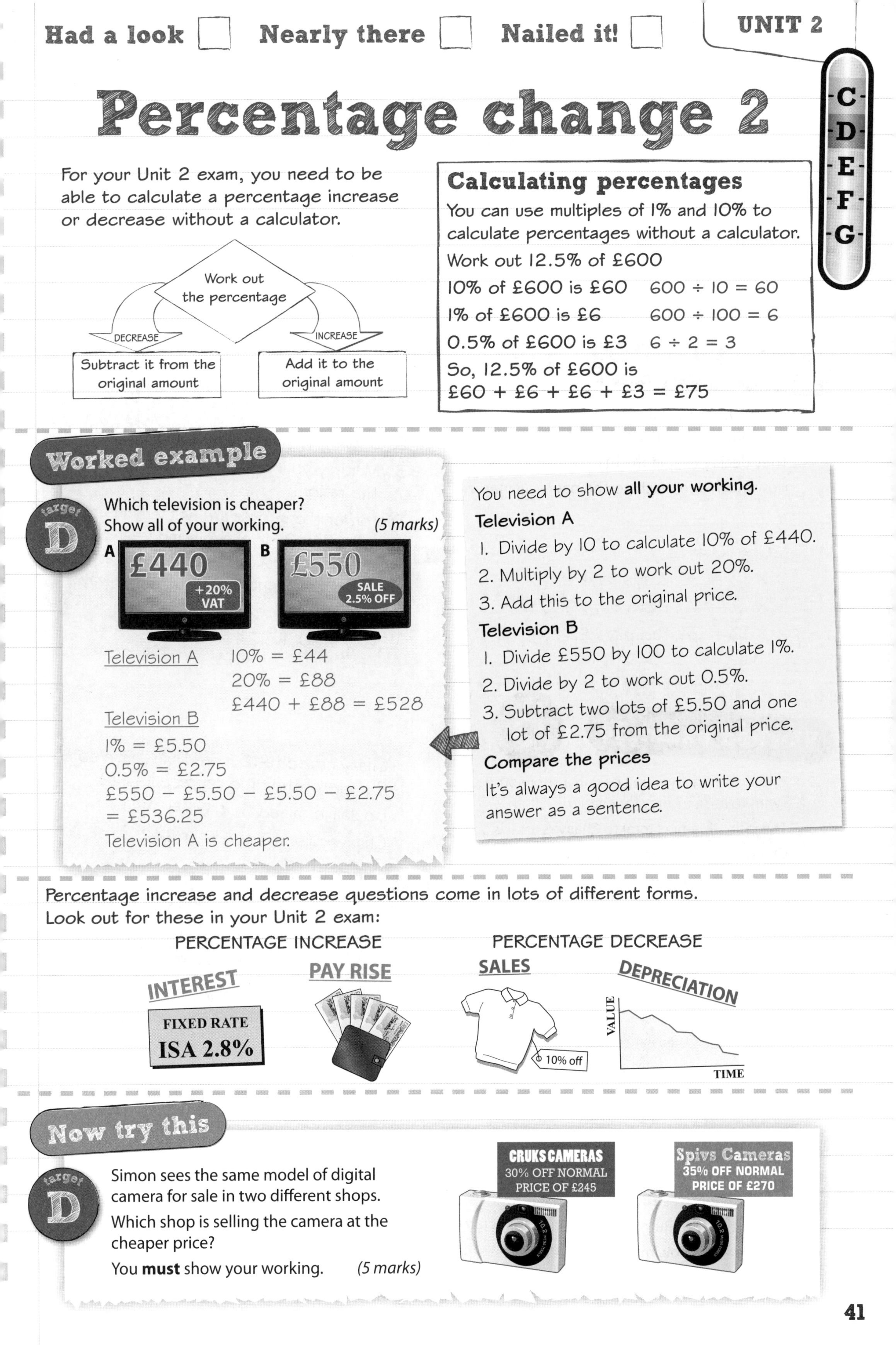

Percentage change 2

For your Unit 2 exam, you need to be able to calculate a percentage increase or decrease without a calculator.

Work out the percentage

DECREASE → Subtract it from the original amount

INCREASE → Add it to the original amount

Calculating percentages

You can use multiples of 1% and 10% to calculate percentages without a calculator.

Work out 12.5% of £600

10% of £600 is £60 — 600 ÷ 10 = 60

1% of £600 is £6 — 600 ÷ 100 = 6

0.5% of £600 is £3 — 6 ÷ 2 = 3

So, 12.5% of £600 is
£60 + £6 + £6 + £3 = £75

Worked example

target D

Which television is cheaper?
Show all of your working. *(5 marks)*

Television A — 10% = £44
20% = £88
£440 + £88 = £528

Television B
1% = £5.50
0.5% = £2.75
£550 − £5.50 − £5.50 − £2.75
= £536.25

Television A is cheaper.

You need to show **all your working.**

Television A

1. Divide by 10 to calculate 10% of £440.
2. Multiply by 2 to work out 20%.
3. Add this to the original price.

Television B

1. Divide £550 by 100 to calculate 1%.
2. Divide by 2 to work out 0.5%.
3. Subtract two lots of £5.50 and one lot of £2.75 from the original price.

Compare the prices

It's always a good idea to write your answer as a sentence.

Percentage increase and decrease questions come in lots of different forms. Look out for these in your Unit 2 exam:

PERCENTAGE INCREASE — PERCENTAGE DECREASE

Now try this

target D

Simon sees the same model of digital camera for sale in two different shops.
Which shop is selling the camera at the cheaper price?
You **must** show your working. *(5 marks)*

C D E F G

Ratio is used in lots of exam questions. Practise these questions WITHOUT a calculator – remember that you won't have one in your Unit 2 exam.

You can answer lots of ratio questions by working out what ONE PART of the ratio represents.

Worked example

Alexis, Nisha and Paul share a flat. One month their phone bill is £120.

They decide to split the bill in the ratio 3 : 5 : 2.

How much does each person pay? *(3 marks)*

$3 + 5 + 2 = 10$

$120 \div 10 = 12$

$3 \times 12 = 36$. Alexis pays £36.

$5 \times 12 = 60$. Nisha pays £60.

$2 \times 12 = 24$. Paul pays £24.

To divide a quantity in a given ratio:

1. Work out the total number of parts in the ratio.
2. Divide the quantity by this total.
3. Multiply your answer by each part of the ratio.

The order the people are written in is the same as the order of the numbers in the ratio. Alexis is first in the list, so 3 parts of the ratio represents the amount she pays.

Check it!

£36 + £60 + £24 = £120 ✓

Worked example

target C

Jamie and Chaaya took part in a sponsored swim to raise money for charity.

The ratio of Jamie's total to Chaaya's total is 5 : 7.

Chaaya raised £12 more than Jamie.

How much money did they raise in total? *(3 marks)*

$7 - 5 = 2$

2 parts = £12 so 1 part = £6

Jamie = £30

Chaaya = £42

Total = £30 + £42 = £72

Chaaya raised £12 more than Jamie so 2 parts of the ratio represents £12.

So Jamie raised 5 × £6 = £30

Chaaya raised 7 × £6 = £42

Check it!

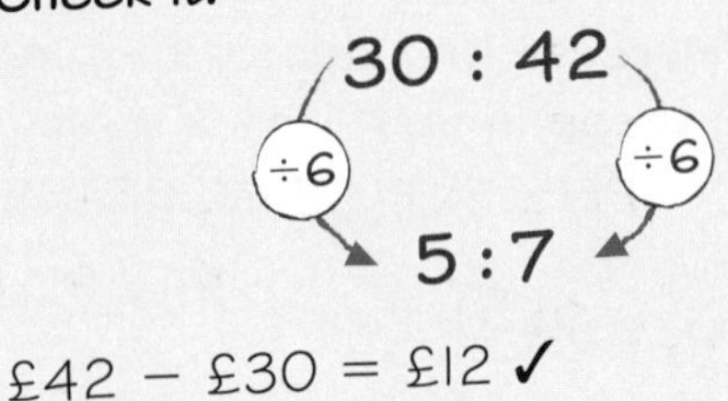

£42 – £30 = £12 ✓

Now try this

Ruth, Sue and Tess share £640 between them.

Ruth receives £80 more than Sue.

The ratio of Ruth's share to Sue's share is 7 : 5

Work out how much Tess receives. *(4 marks)*

2 parts of the ratio represents £80. Use this information to work out how much Sue receives. Remember to show all your working.

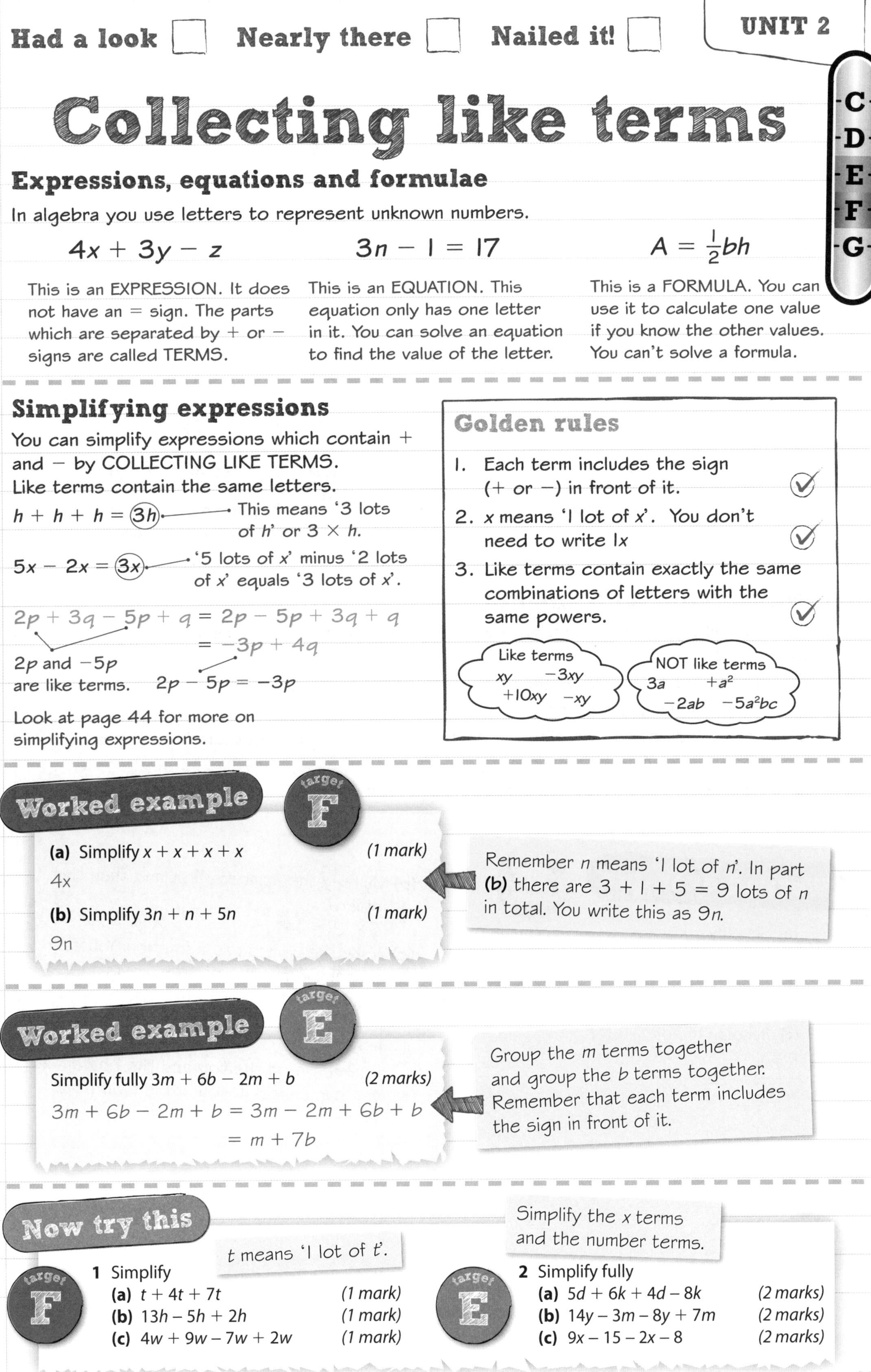

Had a look ☐ Nearly there ☐ Nailed it! ☐

C D E F G

Collecting like terms

Expressions, equations and formulae

In algebra you use letters to represent unknown numbers.

$4x + 3y - z$

This is an EXPRESSION. It does not have an = sign. The parts which are separated by + or − signs are called TERMS.

$3n - 1 = 17$

This is an EQUATION. This equation only has one letter in it. You can solve an equation to find the value of the letter.

$A = \frac{1}{2}bh$

This is a FORMULA. You can use it to calculate one value if you know the other values. You can't solve a formula.

Simplifying expressions

You can simplify expressions which contain + and − by COLLECTING LIKE TERMS.
Like terms contain the same letters.

$h + h + h = 3h$ — This means '3 lots of h' or $3 \times h$.

$5x - 2x = 3x$ — '5 lots of x' minus '2 lots of x' equals '3 lots of x'.

$2p + 3q - 5p + q = 2p - 5p + 3q + q$
$= -3p + 4q$

$2p$ and $-5p$ are like terms. $2p - 5p = -3p$

Look at page 44 for more on simplifying expressions.

Golden rules

1. Each term includes the sign (+ or −) in front of it. ✓
2. x means '1 lot of x'. You don't need to write $1x$ ✓
3. Like terms contain exactly the same combinations of letters with the same powers. ✓

Like terms: xy, $-3xy$, $+10xy$, $-xy$

NOT like terms: $3a$, $+a^2$, $-2ab$, $-5a^2bc$

Worked example (target F)

(a) Simplify $x + x + x + x$ *(1 mark)*

4x

(b) Simplify $3n + n + 5n$ *(1 mark)*

9n

Remember n means '1 lot of n'. In part **(b)** there are $3 + 1 + 5 = 9$ lots of n in total. You write this as $9n$.

Worked example (target E)

Simplify fully $3m + 6b - 2m + b$ *(2 marks)*

$3m + 6b - 2m + b = 3m - 2m + 6b + b$
$= m + 7b$

Group the m terms together and group the b terms together. Remember that each term includes the sign in front of it.

Now try this

t means '1 lot of t'.

1 Simplify (target F)

(a) $t + 4t + 7t$ *(1 mark)*
(b) $13h - 5h + 2h$ *(1 mark)*
(c) $4w + 9w - 7w + 2w$ *(1 mark)*

Simplify the x terms and the number terms.

2 Simplify fully (target E)

(a) $5d + 6k + 4d - 8k$ *(2 marks)*
(b) $14y - 3m - 8y + 7m$ *(2 marks)*
(c) $9x - 15 - 2x - 8$ *(2 marks)*

C D E F G

Simplifying expressions

You'll need to be able to SIMPLIFY expressions which contain × and ÷ in your exam. Use these rules to help you.

1 Multiplying expressions

1. Multiply any number parts first.
2. Then multiply the letters. Remember to use $\square^2$ for letters which are multiplied twice or $\square^3$ for letters which are multiplied three times.

$10a \times 3a = 30a^2$

$10 \times 3 = 30$ $\quad a \times a = a^2$

$3s \times 6t = 18st$

$3 \times 6 = 18$ $\quad s \times t = st$

2 Dividing expressions

1. Write the division as a fraction.
2. Cancel any number parts.
3. If the same letter appears on the top and bottom, you can cancel that as well.

$8y \div 4 = \frac{8y}{4} = 2y$ $\quad 8 \div 4 = 2$

$\frac{36ab}{9b} = 4a$ $\quad 36 \div 9 = 4$

b appears on the top and the bottom, so cancel

Multiplying with algebra

You can multiply letters in algebra by writing them next to each other.
$ab = a \times b$

You can use indices to describe a letter multiplied by itself.
$y \times y = y^2$ You say 'y squared'.

You can use indices to describe the same letter multiplied together three times.
$n \times n \times n = n^3$ You say 'n cubed'.

For a reminder about squares and cubes have a look at page 35.

Worked example

target E

(a) Simplify $a \times a \times a$ *(1 mark)*

a^3

(b) Simplify $7x \times 2y$ *(2 marks)*

$14xy$

(c) Simplify $10pq \div 2p$ *(2 marks)*

$\frac{10pq}{2p} = 5q$

(b) Multiply the numbers first and then the letters.
$7 \times 2 = 14$ $\quad x \times y = xy$

(c) Write the division as a fraction. You can cancel the number parts by dividing top and bottom by 2.
$10 \div 2 = 5$ so write 5 on top of your fraction.
p appears on the top and the bottom so you can cancel it. You are left with $5q$.

Now try this

target E-D

Simplify

(a) $n \times n \times n \times n \times n$ *(1 mark)*

(b) $3r \times 2r \times r$ *(2 marks)*

(c) $4y \times 8z$ *(2 marks)*

(d) $20fg \div 5g$ *(2 marks)*

(e) $28wk \div 4wk$ *(2 marks)*

$20fg \div 5g = \frac{20fg}{5g}$. Cancel any number parts then cancel any letters which are on the top and bottom of the fraction.

C D E F G

Indices

Index notation

These numbers and expressions are written in index notation.

12^3 — the 3 is called the INDEX. The plural of index is indices. The 12 is called the BASE.

n^5 This means $n \times n \times n \times n \times n$.

Index laws

You can use these three index laws to simplify powers and algebraic expressions.

1 To multiply powers of the same base, add the indices.

$a^m \times a^n = a^{m+n}$

$4^3 \times 4^7 = 4^{3+7} = 4^{10}$

$x^4 \times x^3 = x^{4+3} = x^7$

2 To divide powers of the same base, subtract the indices.

$a^m \div a^n = \frac{a^m}{a^n} = a^{m-n}$

$12^8 \div 12^3 = 12^{8-3} = 12^5$

$\frac{m^8}{m^2} = m^{8-2} = m^6$

3 To raise a power of a base to a further power, multiply the indices.

$(a^m)^n = a^{mn}$

$(7^3)^5 = 7^{3 \times 5} = 7^{15}$

$(j^2)^4 = j^{2 \times 4} = j^8$

One at a time

When you are multiplying or dividing expressions with powers:

1. Multiply or divide any number parts first.
2. Use the index laws to work out the new power.

$7x \times 5x^6 = 35x^7$ ($7 \times 5 = 35$; $x \times x^6 = x^{1+6} = x^7$)

$\frac{12a^5}{3a^2} = 4a^3$ ($12 \div 3 = 4$; $a^5 \div a^2 = a^{5-2} = a^3$)

Worked example (target C)

(a) Simplify $p^2 \times p^7$ *(1 mark)*

$p^2 \times p^7 = p^{2+7} = p^9$

(b) Simplify $m^8 \div m^3$ *(1 mark)*

$m^8 \div m^3 = m^{8-3} = m^5$

(c) Simplify $(a^6)^3$ *(1 mark)*

$(a^6)^3 = a^{6 \times 3} = a^{18}$

(a) Add the indices. $a^m \times a^n = a^{m+n}$
(b) Subtract the indices. $a^m \div a^n = a^{m-n}$
(c) Multiply the indices. $(a^m)^n = a^{mn}$

Watch out!

You can only use the index laws when the bases are the same.

If there's no index then the number has a power of 1.

$6^3 \times 6 = 6^{3+1} = 6^4$

$x^8 \div x = x^{8-1} = x^7$

Now try this (target C)

Simplify

(a) $m^5 \times m^4$ *(1 mark)*
(b) $h^{11} \div h^3$ *(1 mark)*
(c) $(y^6)^3$ *(1 mark)*
(d) $\frac{t^{25}}{t^5}$ *(1 mark)*
(e) $k^5 \times k$ *(1 mark)*
(f) $\frac{a^3 \times a^6}{a^4}$ *(1 mark)*

Simplify the top of the fraction first.

C D E F G

Expanding brackets

Expanding brackets is sometimes called MULTIPLYING OUT brackets.

Golden rule

You have to multiply the expression outside the brackets by everything inside the brackets.

$4n \times n = 4n^2$

$$4n(n + 2) = 4n^2 + 8n$$

$4n \times 2 = 8n$

For a reminder about multiplying expressions look back at page 44.

You need to be extra careful if there are negative signs outside the brackets.

$-2 \times x = -2x$

$$-2(x - y) = -2x + 2y$$

$-2 \times -y = 2y$

Multiplying negative terms is like multiplying negative numbers.

When both terms are + or both terms are − the answer is POSITIVE.
$-2a \times -a = +2a^2$

When one term is + and one term is − the answer is NEGATIVE.
$-10p \times 5q = -50pq$

Sometimes you have to EXPAND AND SIMPLIFY. This means 'multiply out the brackets and then collect like terms'.

$6 \times m = 6m$ $\quad 4 \times 3 = 12$

$$6(m + 2) + 4(3 - m) = 6m + 12 + 12 - 4m$$

$6 \times 2 = 12$ $\quad 4 \times -m = -4m$

$$= 2m + 24$$

$6m - 4m = 2m$ $\quad 12 + 12 = 24$

Remember that any negative signs belong to the term on their right.

Look back at page 43 for a reminder about collecting like terms.

Worked example

target C

(a) Expand $5(2y - 3)$ *(2 marks)*

$5(2y - 3) = 10y - 15$

(b) Expand and simplify $2(3x + 4) - 3(4x - 5)$ *(3 marks)*

$2(3x + 4) - 3(4x - 5)$
$= 6x + 8 - 12x + 15$
$= -6x + 23$

For part **(b)**, be careful with the second bracket.
$-3 \times 4x = -12x$
$-3 \times -5 = +15$
The question says 'expand and simplify' so remember to collect any like terms after you have multiplied out the brackets.

Now try this

target D

1 Expand **(a)** $3(y - 6)$ *(2 marks)* **(b)** $m(m + 7)$ *(2 marks)*

target C

2 Expand and simplify

(a) $5(a - 2b) + 4(2a + b)$ *(3 marks)*

(b) $2(4w + 3) - 3(3w - 5)$ *(3 marks)*

(c) $2m(m + 9) - m(m - 4)$ *(3 marks)*

Be careful when there is a minus sign in front of the second bracket. $-3 \times 3w = -9w$ and $-3 \times -5 = 15$.

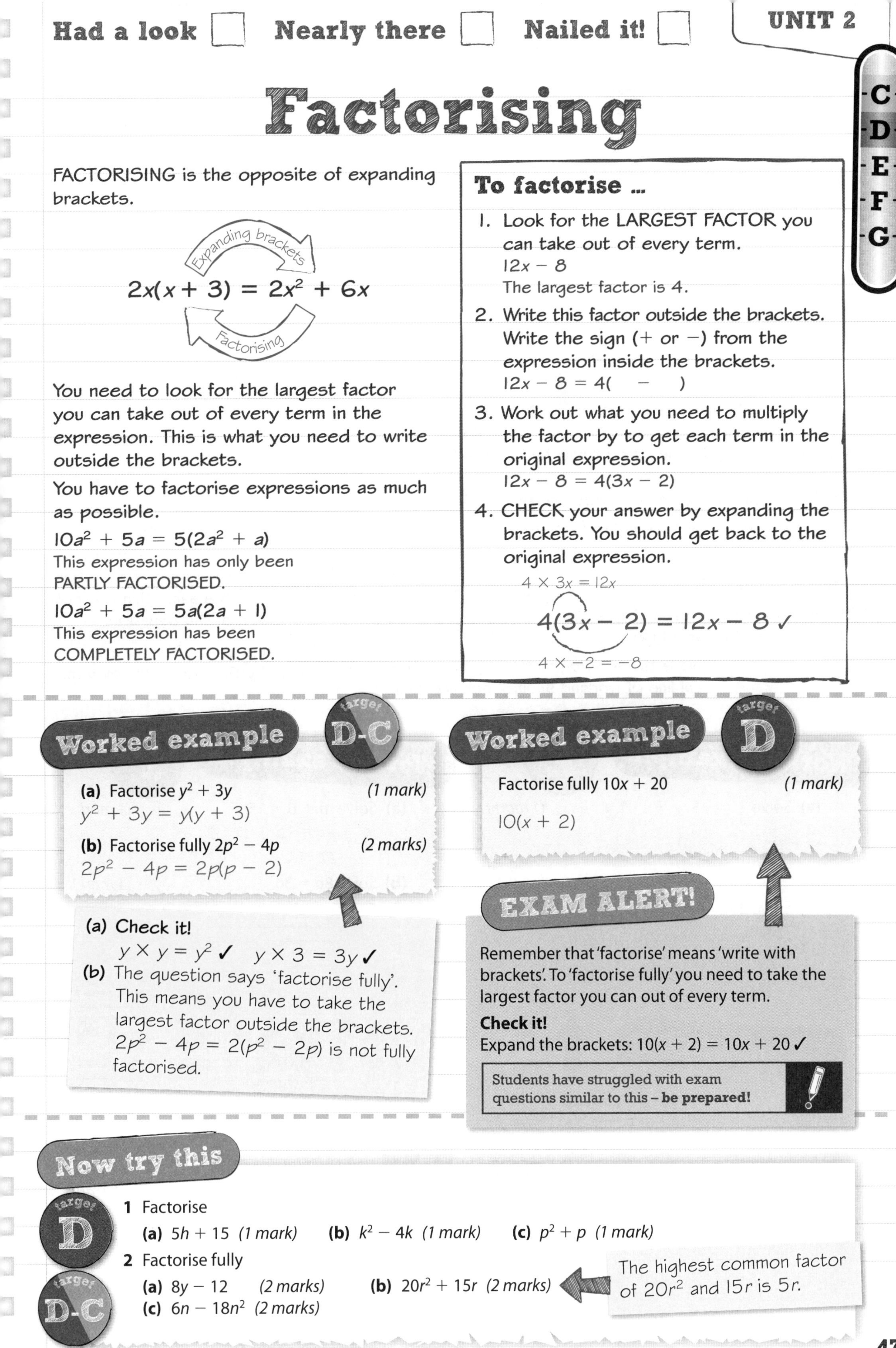

Factorising

FACTORISING is the opposite of expanding brackets.

You need to look for the largest factor you can take out of every term in the expression. This is what you need to write outside the brackets.

You have to factorise expressions as much as possible.

$10a^2 + 5a = 5(2a^2 + a)$

This expression has only been PARTLY FACTORISED.

$10a^2 + 5a = 5a(2a + 1)$

This expression has been COMPLETELY FACTORISED.

To factorise ...

1. Look for the LARGEST FACTOR you can take out of every term.
   $12x - 8$
   The largest factor is 4.
2. Write this factor outside the brackets. Write the sign (+ or −) from the expression inside the brackets.
   $12x - 8 = 4(\quad - \quad)$
3. Work out what you need to multiply the factor by to get each term in the original expression.
   $12x - 8 = 4(3x - 2)$
4. CHECK your answer by expanding the brackets. You should get back to the original expression.

$4 \times 3x = 12x$

$4(3x - 2) = 12x - 8$ ✓

$4 \times -2 = -8$

Worked example (Target D-C)

(a) Factorise $y^2 + 3y$ *(1 mark)*

$y^2 + 3y = y(y + 3)$

(b) Factorise fully $2p^2 - 4p$ *(2 marks)*

$2p^2 - 4p = 2p(p - 2)$

(a) Check it!
$y \times y = y^2$ ✓ $y \times 3 = 3y$ ✓

(b) The question says 'factorise fully'. This means you have to take the largest factor outside the brackets. $2p^2 - 4p = 2(p^2 - 2p)$ is not fully factorised.

Worked example (Target D)

Factorise fully $10x + 20$ *(1 mark)*

$10(x + 2)$

EXAM ALERT!

Remember that 'factorise' means 'write with brackets'. To 'factorise fully' you need to take the largest factor you can out of every term.

Check it!

Expand the brackets: $10(x + 2) = 10x + 20$ ✓

Students have struggled with exam questions similar to this – **be prepared!**

Now try this

(Target D)

1 Factorise

(a) $5h + 15$ *(1 mark)* **(b)** $k^2 - 4k$ *(1 mark)* **(c)** $p^2 + p$ *(1 mark)*

(Target D-C)

2 Factorise fully

(a) $8y - 12$ *(2 marks)* **(b)** $20r^2 + 15r$ *(2 marks)*

(c) $6n - 18n^2$ *(2 marks)*

The highest common factor of $20r^2$ and $15r$ is $5r$.

C D E F G

Equations 1

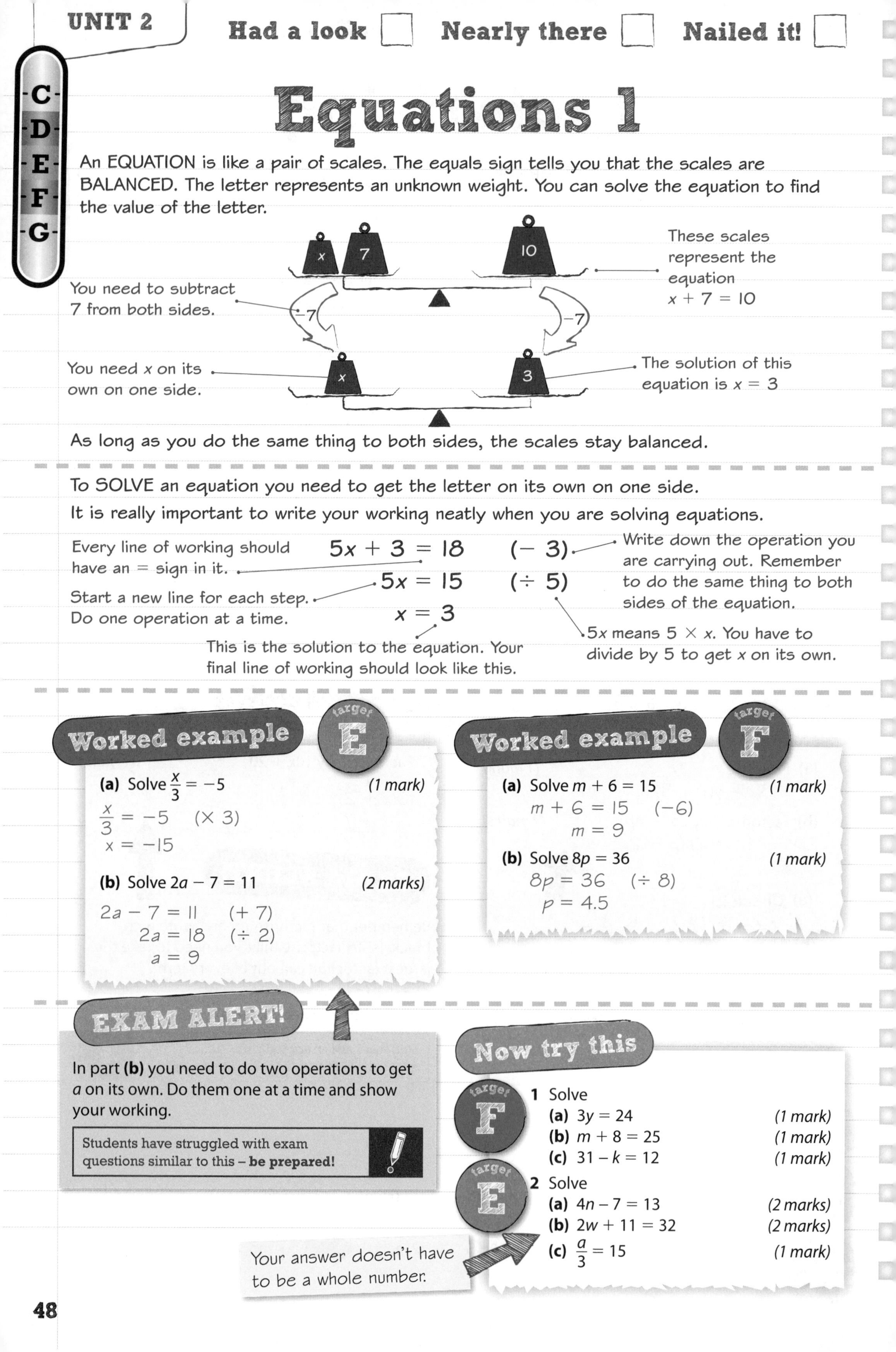

An EQUATION is like a pair of scales. The equals sign tells you that the scales are BALANCED. The letter represents an unknown weight. You can solve the equation to find the value of the letter.

As long as you do the same thing to both sides, the scales stay balanced.

To SOLVE an equation you need to get the letter on its own on one side.

It is really important to write your working neatly when you are solving equations.

$$5x + 3 = 18 \quad (-3)$$
$$5x = 15 \quad (\div 5)$$
$$x = 3$$

Every line of working should have an = sign in it.

Start a new line for each step. Do one operation at a time.

Write down the operation you are carrying out. Remember to do the same thing to both sides of the equation.

$5x$ means $5 \times x$. You have to divide by 5 to get x on its own.

This is the solution to the equation. Your final line of working should look like this.

Worked example (target E)

(a) Solve $\frac{x}{3} = -5$ *(1 mark)*

$\frac{x}{3} = -5 \quad (\times 3)$

$x = -15$

(b) Solve $2a - 7 = 11$ *(2 marks)*

$2a - 7 = 11 \quad (+7)$

$2a = 18 \quad (\div 2)$

$a = 9$

Worked example (target F)

(a) Solve $m + 6 = 15$ *(1 mark)*

$m + 6 = 15 \quad (-6)$

$m = 9$

(b) Solve $8p = 36$ *(1 mark)*

$8p = 36 \quad (\div 8)$

$p = 4.5$

EXAM ALERT!

In part **(b)** you need to do two operations to get *a* on its own. Do them one at a time and show your working.

Students have struggled with exam questions similar to this – **be prepared!**

Now try this

1 Solve (target F)

(a) $3y = 24$ *(1 mark)*

(b) $m + 8 = 25$ *(1 mark)*

(c) $31 - k = 12$ *(1 mark)*

2 Solve (target E)

(a) $4n - 7 = 13$ *(2 marks)*

(b) $2w + 11 = 32$ *(2 marks)*

(c) $\frac{a}{3} = 15$ *(1 mark)*

Your answer doesn't have to be a whole number.

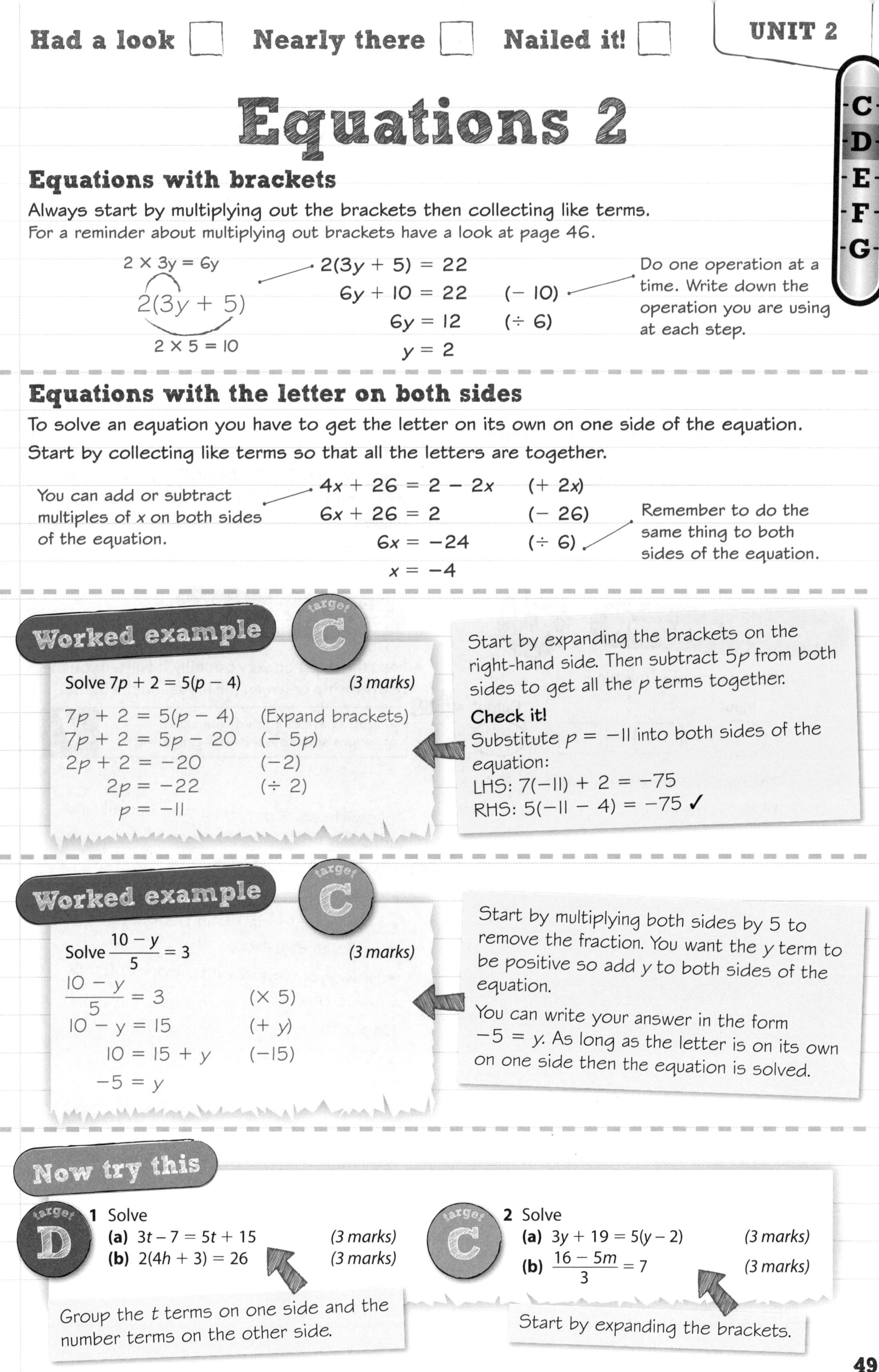

Equations 2

Equations with brackets

Always start by multiplying out the brackets then collecting like terms.
For a reminder about multiplying out brackets have a look at page 46.

$2 \times 3y = 6y$

$2(3y + 5)$

$2 \times 5 = 10$

$$
\begin{aligned}
2(3y + 5) &= 22 \\
6y + 10 &= 22 \quad (-10) \\
6y &= 12 \quad (\div 6) \\
y &= 2
\end{aligned}
$$

Do one operation at a time. Write down the operation you are using at each step.

Equations with the letter on both sides

To solve an equation you have to get the letter on its own on one side of the equation.
Start by collecting like terms so that all the letters are together.

You can add or subtract multiples of x on both sides of the equation.

$$
\begin{aligned}
4x + 26 &= 2 - 2x \quad (+2x) \\
6x + 26 &= 2 \quad (-26) \\
6x &= -24 \quad (\div 6) \\
x &= -4
\end{aligned}
$$

Remember to do the same thing to both sides of the equation.

Worked example (target C)

Solve $7p + 2 = 5(p - 4)$ *(3 marks)*

$$
\begin{aligned}
7p + 2 &= 5(p - 4) \quad \text{(Expand brackets)} \\
7p + 2 &= 5p - 20 \quad (-5p) \\
2p + 2 &= -20 \quad (-2) \\
2p &= -22 \quad (\div 2) \\
p &= -11
\end{aligned}
$$

Start by expanding the brackets on the right-hand side. Then subtract $5p$ from both sides to get all the p terms together.

Check it!
Substitute $p = -11$ into both sides of the equation:
LHS: $7(-11) + 2 = -75$
RHS: $5(-11 - 4) = -75$ ✓

Worked example (target C)

Solve $\frac{10 - y}{5} = 3$ *(3 marks)*

$$
\begin{aligned}
\frac{10 - y}{5} &= 3 \quad (\times 5) \\
10 - y &= 15 \quad (+y) \\
10 &= 15 + y \quad (-15) \\
-5 &= y
\end{aligned}
$$

Start by multiplying both sides by 5 to remove the fraction. You want the y term to be positive so add y to both sides of the equation.

You can write your answer in the form $-5 = y$. As long as the letter is on its own on one side then the equation is solved.

Now try this

Target D

1 Solve
(a) $3t - 7 = 5t + 15$ *(3 marks)*
(b) $2(4h + 3) = 26$ *(3 marks)*

Group the t terms on one side and the number terms on the other side.

Target C

2 Solve
(a) $3y + 19 = 5(y - 2)$ *(3 marks)*
(b) $\frac{16 - 5m}{3} = 7$ *(3 marks)*

Start by expanding the brackets.

Had a look ☐ Nearly there ☐ Nailed it! ☐

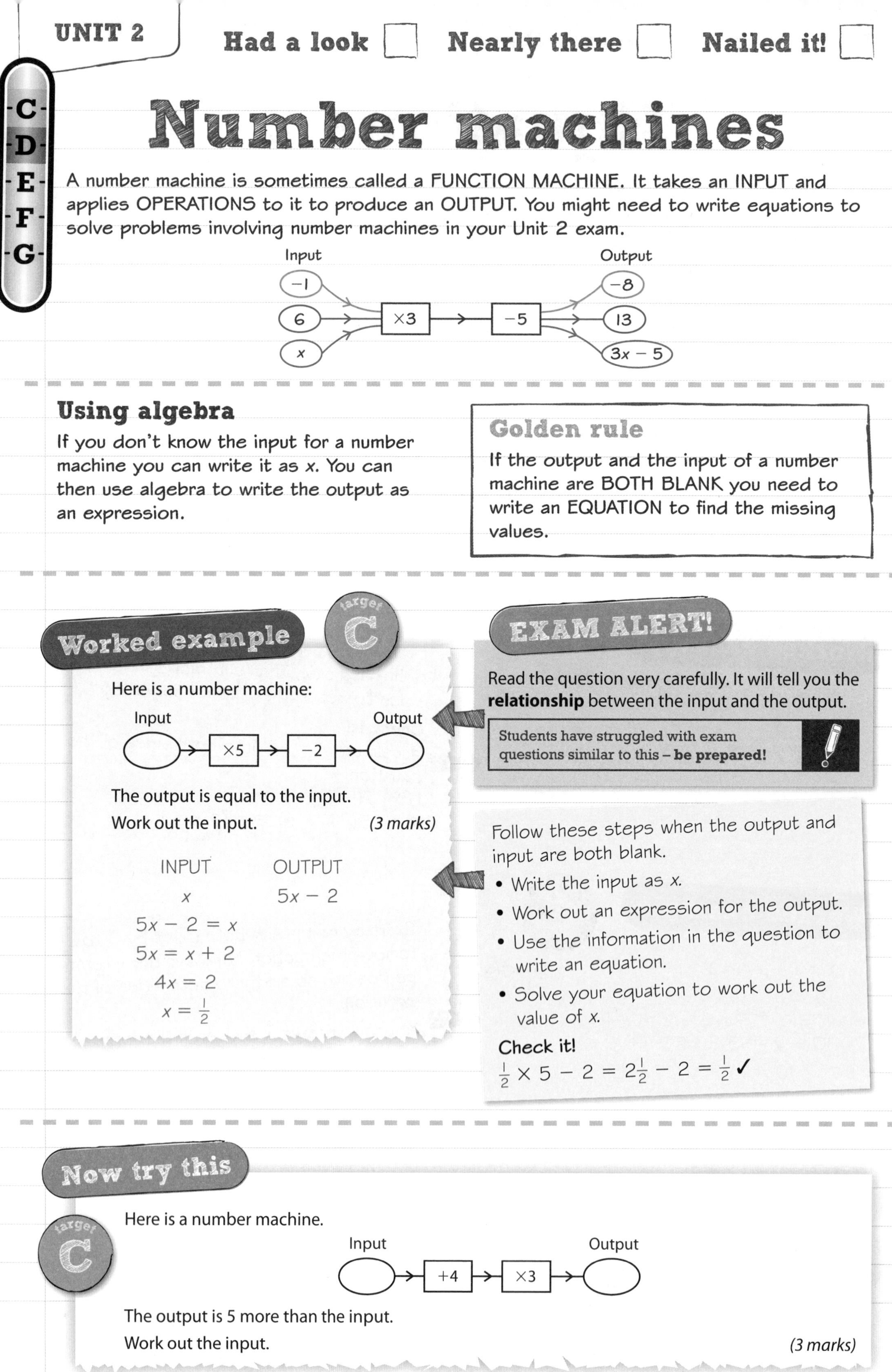

Number machines

A number machine is sometimes called a FUNCTION MACHINE. It takes an INPUT and applies OPERATIONS to it to produce an OUTPUT. You might need to write equations to solve problems involving number machines in your Unit 2 exam.

Using algebra

If you don't know the input for a number machine you can write it as x. You can then use algebra to write the output as an expression.

Golden rule

If the output and the input of a number machine are BOTH BLANK you need to write an EQUATION to find the missing values.

Worked example

target C

Here is a number machine:

The output is equal to the input.

Work out the input. *(3 marks)*

INPUT OUTPUT

x $5x - 2$

$5x - 2 = x$

$5x = x + 2$

$4x = 2$

$x = \frac{1}{2}$

EXAM ALERT!

Read the question very carefully. It will tell you the **relationship** between the input and the output.

Students have struggled with exam questions similar to this – **be prepared!**

Follow these steps when the output and input are both blank.

- Write the input as x.
- Work out an expression for the output.
- Use the information in the question to write an equation.
- Solve your equation to work out the value of x.

Check it!

$\frac{1}{2} \times 5 - 2 = 2\frac{1}{2} - 2 = \frac{1}{2}$ ✓

Now try this

target C

Here is a number machine.

The output is 5 more than the input.

Work out the input. *(3 marks)*

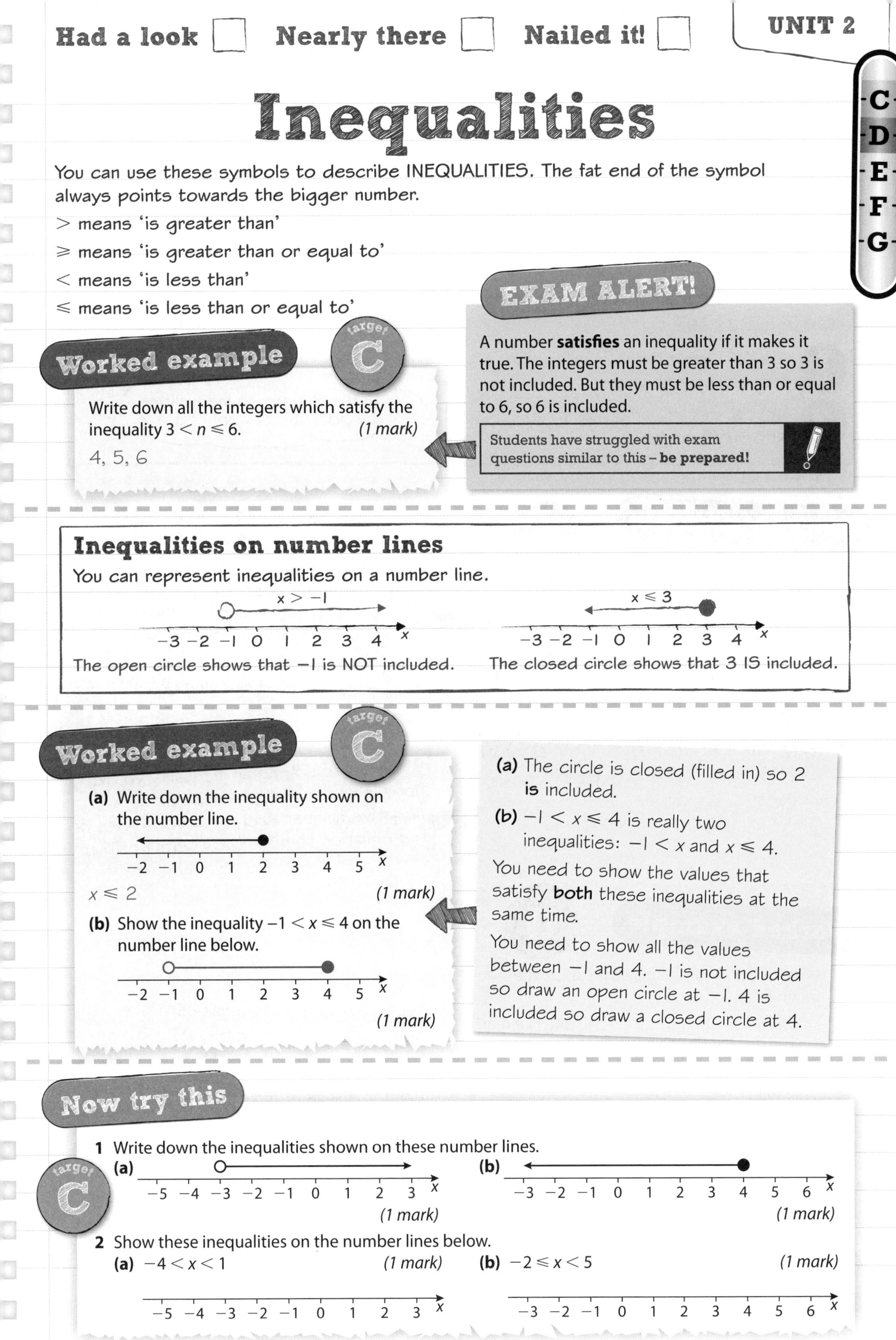

Inequalities

You can use these symbols to describe INEQUALITIES. The fat end of the symbol always points towards the bigger number.

$>$ means 'is greater than'

$\geqslant$ means 'is greater than or equal to'

$<$ means 'is less than'

$\leqslant$ means 'is less than or equal to'

Worked example

target C

Write down all the integers which satisfy the inequality $3 < n \leqslant 6$. *(1 mark)*

4, 5, 6

EXAM ALERT!

A number **satisfies** an inequality if it makes it true. The integers must be greater than 3 so 3 is not included. But they must be less than or equal to 6, so 6 is included.

Students have struggled with exam questions similar to this – **be prepared!**

Inequalities on number lines

You can represent inequalities on a number line.

$x > -1$

The open circle shows that -1 is NOT included.

$x \leqslant 3$

The closed circle shows that 3 IS included.

Worked example

target C

(a) Write down the inequality shown on the number line.

$x \leqslant 2$ *(1 mark)*

(b) Show the inequality $-1 < x \leqslant 4$ on the number line below.

(1 mark)

(a) The circle is closed (filled in) so 2 **is** included.

(b) $-1 < x \leqslant 4$ is really two inequalities: $-1 < x$ and $x \leqslant 4$.

You need to show the values that satisfy **both** these inequalities at the same time.

You need to show all the values between -1 and 4. -1 is not included so draw an open circle at -1. 4 is included so draw a closed circle at 4.

Now try this

target C

1 Write down the inequalities shown on these number lines.

(a) *(1 mark)*

(b) *(1 mark)*

2 Show these inequalities on the number lines below.

(a) $-4 < x < 1$ *(1 mark)*

(b) $-2 \leqslant x < 5$ *(1 mark)*

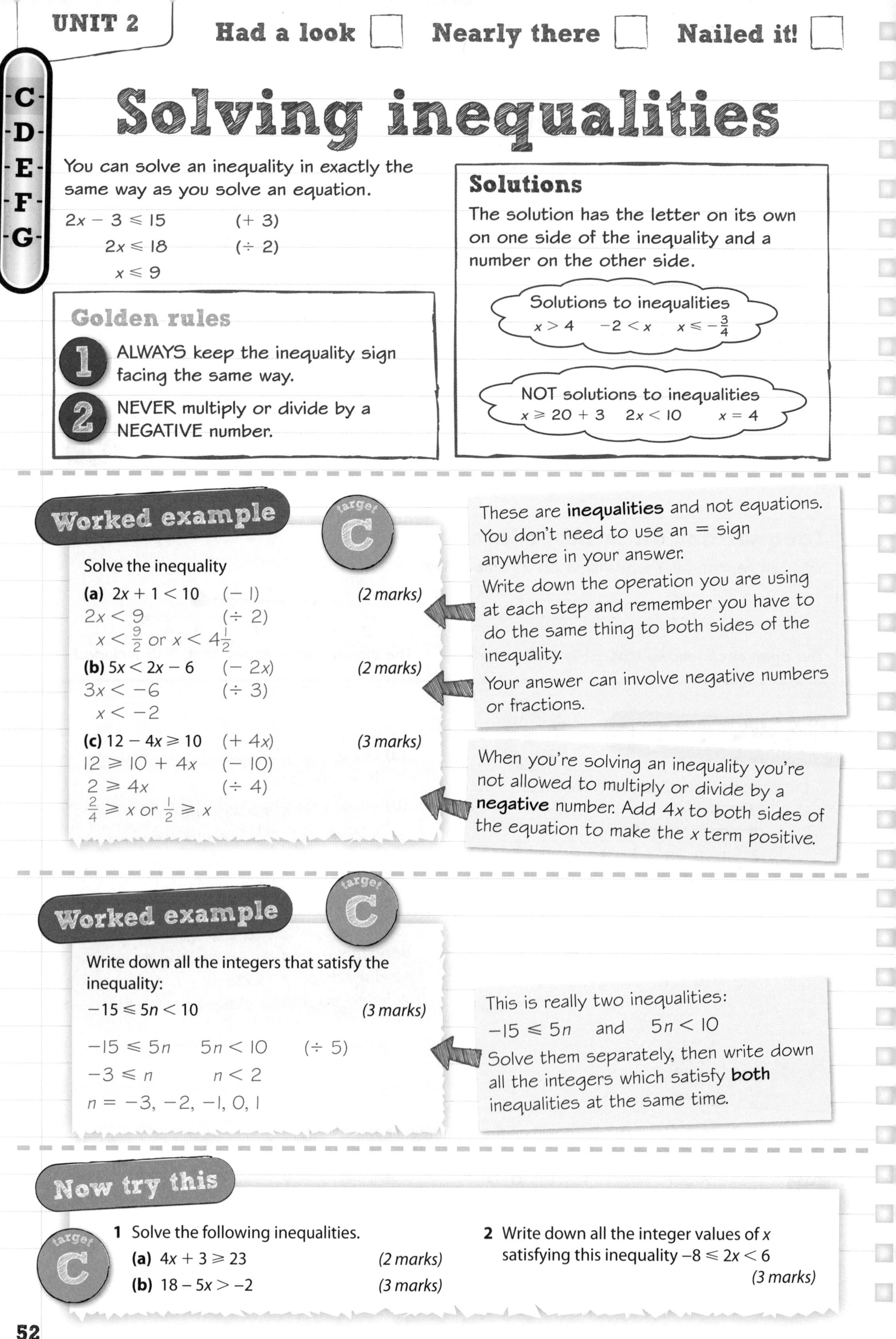

C D E F G

Solving inequalities

You can solve an inequality in exactly the same way as you solve an equation.

$2x - 3 \leqslant 15$ (+ 3)

$2x \leqslant 18$ (÷ 2)

$x \leqslant 9$

Golden rules

1. ALWAYS keep the inequality sign facing the same way.
2. NEVER multiply or divide by a NEGATIVE number.

Solutions

The solution has the letter on its own on one side of the inequality and a number on the other side.

Solutions to inequalities

$x > 4$ $-2 < x$ $x \leqslant -\frac{3}{4}$

NOT solutions to inequalities

$x \geqslant 20 + 3$ $2x < 10$ $x = 4$

Worked example

target C

Solve the inequality

(a) $2x + 1 < 10$ (− 1) *(2 marks)*

$2x < 9$ (÷ 2)

$x < \frac{9}{2}$ or $x < 4\frac{1}{2}$

(b) $5x < 2x - 6$ (− 2x) *(2 marks)*

$3x < -6$ (÷ 3)

$x < -2$

(c) $12 - 4x \geqslant 10$ (+ 4x) *(3 marks)*

$12 \geqslant 10 + 4x$ (− 10)

$2 \geqslant 4x$ (÷ 4)

$\frac{2}{4} \geqslant x$ or $\frac{1}{2} \geqslant x$

These are **inequalities** and not equations. You don't need to use an = sign anywhere in your answer.

Write down the operation you are using at each step and remember you have to do the same thing to both sides of the inequality.

Your answer can involve negative numbers or fractions.

When you're solving an inequality you're not allowed to multiply or divide by a **negative** number. Add $4x$ to both sides of the equation to make the x term positive.

Worked example

target C

Write down all the integers that satisfy the inequality:

$-15 \leqslant 5n < 10$ *(3 marks)*

$-15 \leqslant 5n$ $5n < 10$ (÷ 5)

$-3 \leqslant n$ $n < 2$

$n = -3, -2, -1, 0, 1$

This is really two inequalities:

$-15 \leqslant 5n$ and $5n < 10$

Solve them separately, then write down all the integers which satisfy **both** inequalities at the same time.

Now try this

target C

1 Solve the following inequalities.

(a) $4x + 3 \geqslant 23$ *(2 marks)*

(b) $18 - 5x > -2$ *(3 marks)*

2 Write down all the integer values of x satisfying this inequality $-8 \leqslant 2x < 6$ *(3 marks)*

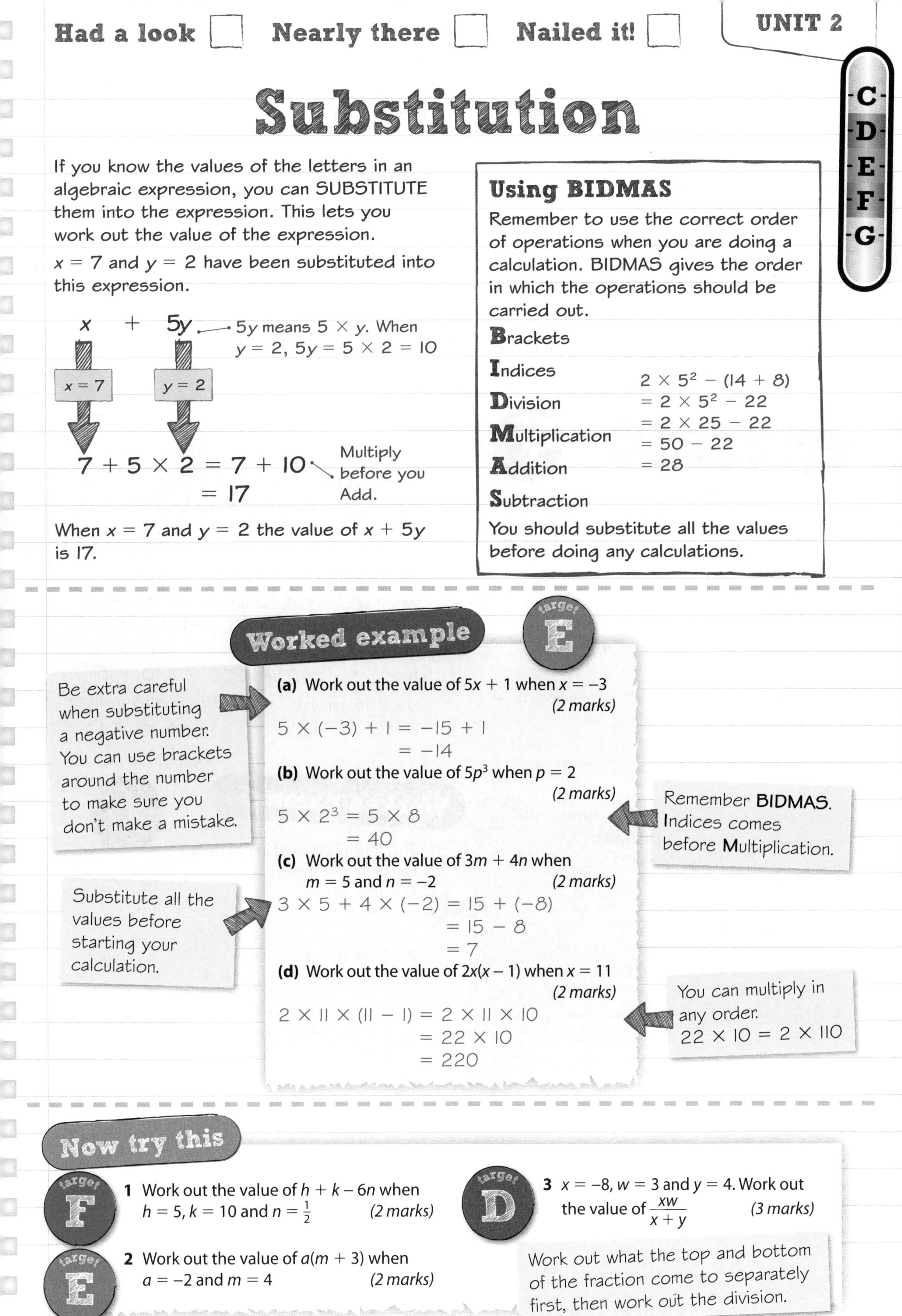

Substitution

If you know the values of the letters in an algebraic expression, you can SUBSTITUTE them into the expression. This lets you work out the value of the expression.

$x = 7$ and $y = 2$ have been substituted into this expression.

$x + 5y$ — $5y$ means $5 \times y$. When $y = 2$, $5y = 5 \times 2 = 10$

$x = 7$ $y = 2$

$7 + 5 \times 2 = 7 + 10$ — Multiply before you Add.

$= 17$

When $x = 7$ and $y = 2$ the value of $x + 5y$ is 17.

Using BIDMAS

Remember to use the correct order of operations when you are doing a calculation. BIDMAS gives the order in which the operations should be carried out.

Brackets
Indices
Division
Multiplication
Addition
Subtraction

$2 \times 5^2 - (14 + 8)$
$= 2 \times 5^2 - 22$
$= 2 \times 25 - 22$
$= 50 - 22$
$= 28$

You should substitute all the values before doing any calculations.

Worked example (target E)

Be extra careful when substituting a negative number. You can use brackets around the number to make sure you don't make a mistake.

(a) Work out the value of $5x + 1$ when $x = -3$ *(2 marks)*

$5 \times (-3) + 1 = -15 + 1$
$= -14$

(b) Work out the value of $5p^3$ when $p = 2$ *(2 marks)*

$5 \times 2^3 = 5 \times 8$
$= 40$

Remember BIDMAS. Indices comes before Multiplication.

(c) Work out the value of $3m + 4n$ when $m = 5$ and $n = -2$ *(2 marks)*

$3 \times 5 + 4 \times (-2) = 15 + (-8)$
$= 15 - 8$
$= 7$

Substitute all the values before starting your calculation.

(d) Work out the value of $2x(x - 1)$ when $x = 11$ *(2 marks)*

$2 \times 11 \times (11 - 1) = 2 \times 11 \times 10$
$= 22 \times 10$
$= 220$

You can multiply in any order. $22 \times 10 = 2 \times 110$

Now try this

(target F) **1** Work out the value of $h + k - 6n$ when $h = 5$, $k = 10$ and $n = \frac{1}{2}$ *(2 marks)*

(target E) **2** Work out the value of $a(m + 3)$ when $a = -2$ and $m = 4$ *(2 marks)*

(target D) **3** $x = -8$, $w = 3$ and $y = 4$. Work out the value of $\frac{xw}{x + y}$ *(3 marks)*

Work out what the top and bottom of the fraction come to separately first, then work out the division.

Formulae

A FORMULA is a mathematical rule.
Formulae is the plural of formula.
The formula for the area of this triangle is:
Area = $\frac{1}{2}$ × base × height
You can write this formula using algebra as:
$A = \frac{1}{2}bh$

height
base

A formula lets you calculate one quantity when you know the others.
You need to SUBSTITUTE the values you know into the formula.
For more on substitution have a look at page 53.

Worked example (target F)

You can use this formula to work out the cooking time in minutes for a turkey.

Cooking time = Weight in kg × 30 + 45

Work out the cooking time for a turkey weighing 7 kg. *(3 marks)*

Cooking time = 7 × 30 + 45
= 210 + 45
= 255

The cooking time is 255 minutes, or 4 hours and 15 minutes.

Substitute the weight into your formula before you do any calculations.
Remember to use **BIDMAS** for the correct order of operations. You **M**ultiply before you **A**dd. For a reminder about substituting and BIDMAS have a look at page 53.
When you are giving an answer in minutes and it is larger than 60 minutes, you can give it in minutes or in hours and minutes.
4 × 60 = 240 and 255 − 240 = 15
So 255 minutes = 4 hours and 15 minutes.

Worked example (target D)

This formula is used in physics to calculate distance.
$D = ut - 5t^2$
$u = 20$
$t = 3$
Work out the value of D. *(3 marks)*

$D = 20 \times 3 - 5 \times 3^2$
$= 20 \times 3 - 5 \times 9$
$= 60 - 45$
$= 15$

Substitute the values for u and t into the formula. Using **BIDMAS**:
1. Do the **I**ndex (power) first.
   $3^2 = 9$
2. Do the **M**ultiplications next.
   $20 \times 3 = 60$
   $5 \times 9 = 45$
3. Do the **S**ubtraction last.
   $60 - 45 = 15$

Don't try to do more than one operation on each line of working.

Now try this

1 Here is a trapezium.
The area of the trapezium is given by the formula Area = $\frac{1}{2}(a + b)h$
Work out the area when $h = 4.5$, $a = 6.5$ and $b = 9.5$. *(3 marks)*

a
h
b

2 Here is a formula: $D = b^2 - 4ac$
Work out the value of D when $a = 2$, $b = 5$ and $c = -3$. *(3 marks)*

Substitute all the values before you do any calculations.

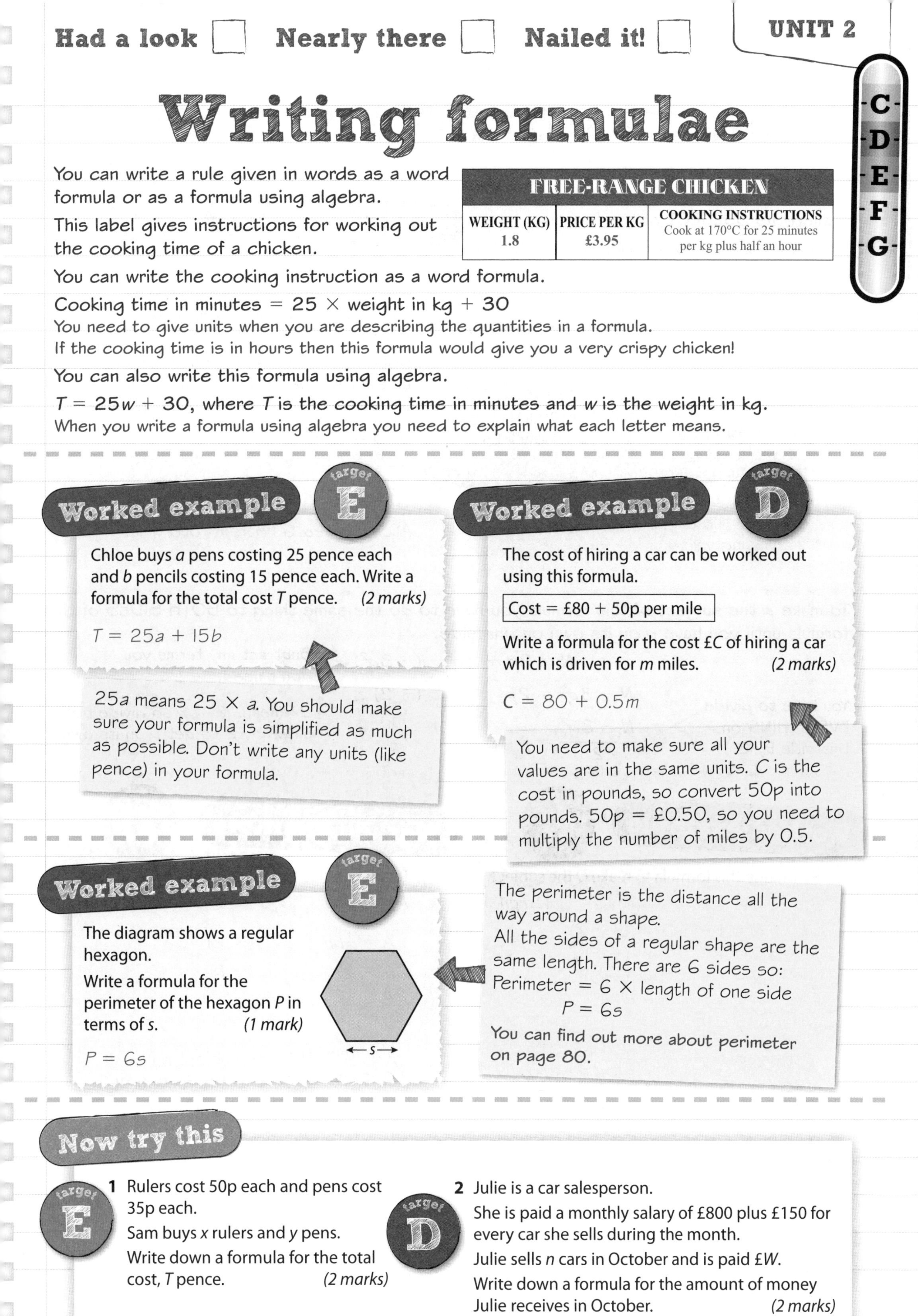

Writing formulae

You can write a rule given in words as a word formula or as a formula using algebra.

This label gives instructions for working out the cooking time of a chicken.

FREE-RANGE CHICKEN		
WEIGHT (KG) 1.8	PRICE PER KG £3.95	COOKING INSTRUCTIONS Cook at 170°C for 25 minutes per kg plus half an hour

You can write the cooking instruction as a word formula.

Cooking time in minutes = 25 × weight in kg + 30

You need to give units when you are describing the quantities in a formula.
If the cooking time is in hours then this formula would give you a very crispy chicken!

You can also write this formula using algebra.

$T = 25w + 30$, where T is the cooking time in minutes and w is the weight in kg.

When you write a formula using algebra you need to explain what each letter means.

Worked example (target E)

Chloe buys a pens costing 25 pence each and b pencils costing 15 pence each. Write a formula for the total cost T pence. *(2 marks)*

$T = 25a + 15b$

$25a$ means $25 \times a$. You should make sure your formula is simplified as much as possible. Don't write any units (like pence) in your formula.

Worked example (target D)

The cost of hiring a car can be worked out using this formula.

Cost = £80 + 50p per mile

Write a formula for the cost £C of hiring a car which is driven for m miles. *(2 marks)*

$C = 80 + 0.5m$

You need to make sure all your values are in the same units. C is the cost in pounds, so convert 50p into pounds. 50p = £0.50, so you need to multiply the number of miles by 0.5.

Worked example (target E)

The diagram shows a regular hexagon.

Write a formula for the perimeter of the hexagon P in terms of s. *(1 mark)*

s

$P = 6s$

The perimeter is the distance all the way around a shape.
All the sides of a regular shape are the same length. There are 6 sides so:
Perimeter = 6 × length of one side
$P = 6s$

You can find out more about perimeter on page 80.

Now try this

1 (target E) Rulers cost 50p each and pens cost 35p each.
Sam buys x rulers and y pens.
Write down a formula for the total cost, T pence. *(2 marks)*

2 (target D) Julie is a car salesperson.
She is paid a monthly salary of £800 plus £150 for every car she sells during the month.
Julie sells n cars in October and is paid £W.
Write down a formula for the amount of money Julie receives in October. *(2 marks)*

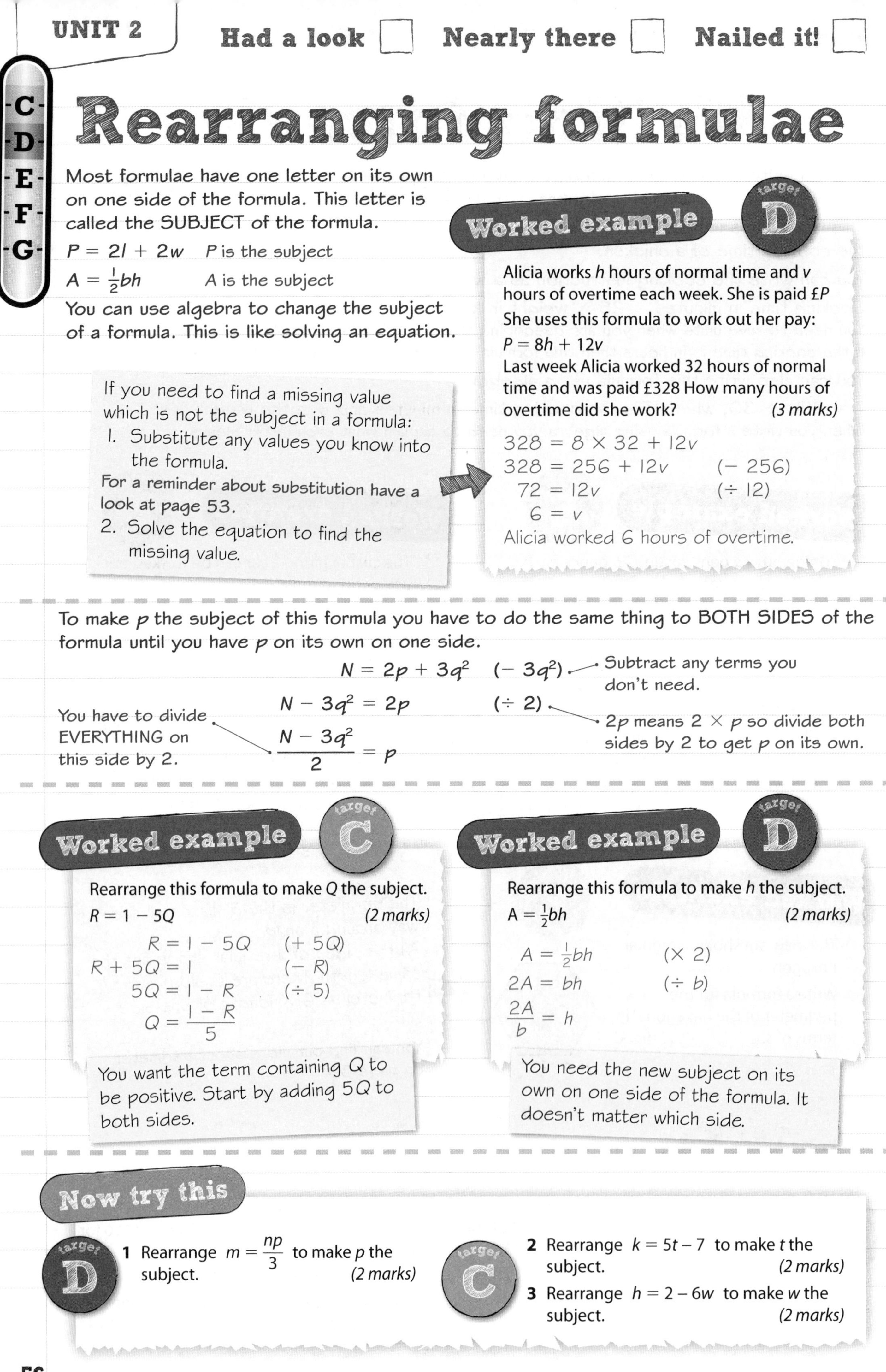

C D E F G

Rearranging formulae

Most formulae have one letter on its own on one side of the formula. This letter is called the SUBJECT of the formula.

$P = 2l + 2w$ P is the subject

$A = \frac{1}{2}bh$ A is the subject

You can use algebra to change the subject of a formula. This is like solving an equation.

If you need to find a missing value which is not the subject in a formula:

1. Substitute any values you know into the formula.

For a reminder about substitution have a look at page 53.

2. Solve the equation to find the missing value.

Worked example (target D)

Alicia works h hours of normal time and v hours of overtime each week. She is paid £P She uses this formula to work out her pay:

$P = 8h + 12v$

Last week Alicia worked 32 hours of normal time and was paid £328 How many hours of overtime did she work? *(3 marks)*

$328 = 8 \times 32 + 12v$

$328 = 256 + 12v$ $(-\ 256)$

$72 = 12v$ $(\div\ 12)$

$6 = v$

Alicia worked 6 hours of overtime.

To make p the subject of this formula you have to do the same thing to BOTH SIDES of the formula until you have p on its own on one side.

$N = 2p + 3q^2$ $(-\ 3q^2)$ — Subtract any terms you don't need.

$N - 3q^2 = 2p$ $(\div\ 2)$ — $2p$ means $2 \times p$ so divide both sides by 2 to get p on its own.

$\dfrac{N - 3q^2}{2} = p$ — You have to divide EVERYTHING on this side by 2.

Worked example (target C)

Rearrange this formula to make Q the subject.

$R = 1 - 5Q$ *(2 marks)*

$R = 1 - 5Q$ $(+\ 5Q)$

$R + 5Q = 1$ $(-\ R)$

$5Q = 1 - R$ $(\div\ 5)$

$Q = \dfrac{1 - R}{5}$

You want the term containing Q to be positive. Start by adding $5Q$ to both sides.

Worked example (target D)

Rearrange this formula to make h the subject.

$A = \frac{1}{2}bh$ *(2 marks)*

$A = \frac{1}{2}bh$ $(\times\ 2)$

$2A = bh$ $(\div\ b)$

$\dfrac{2A}{b} = h$

You need the new subject on its own on one side of the formula. It doesn't matter which side.

Now try this

(target D)

1 Rearrange $m = \dfrac{np}{3}$ to make p the subject. *(2 marks)*

(target C)

2 Rearrange $k = 5t - 7$ to make t the subject. *(2 marks)*

3 Rearrange $h = 2 - 6w$ to make w the subject. *(2 marks)*

Using algebra

You might need to use the information given in a question to write an equation. You might need to choose a letter to represent an unknown quantity you are trying to find. For a reminder about solving equations have a look at pages 48 and 49.

Worked example

Here are three number cards.

Each card has a whole number written on the back.

The number on card B is twice the number on card A.

The number on card C is five more than the number on card A.

The sum of the numbers on all three cards is 37.

Work out the number on each card. *(4 marks)*

$B = 2A \qquad C = A + 5$

$A + B + C = 37$

$A + 2A + A + 5 = 37$

$4A + 5 = 37 \quad (-5)$

$4A = 32 \quad (\div 4)$

$A = 8$

$B = 2 \times 8 = 16 \qquad C = 8 + 5 = 13$

You could try guessing the numbers on the cards, but it might take a long time. You can solve this problem quickly by writing your own equation.

- Use A to represent the number on card A.
- Write expressions for the numbers on cards B and C in terms of A.
- Add together these three expressions and set the total equal to 37. This gives you an equation which you can solve to find the value of A.
- Simplify the left-hand side of the equation by collecting like terms.

Once you have calculated A, remember to work out the values of B and C as well.

Check it!

$8 + 16 + 13 = 37$ ✓

True or false?

You might need to SUBSTITUTE values into an algebraic statement to show that it is FALSE. You only need to find one COUNTER-EXAMPLE to show that a statement is false.

Try some different values of n and write down all your working. You need to be able to recognise all the prime numbers less than 20.

Worked example

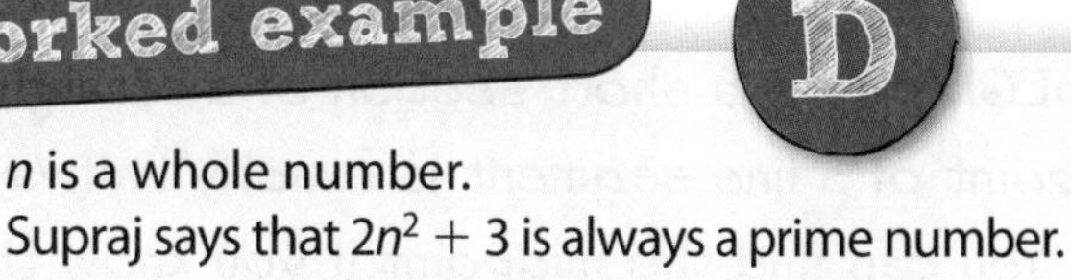

n is a whole number.

Supraj says that $2n^2 + 3$ is always a prime number.

Give an example to show that he is wrong. *(2 marks)*

$n = 1{:}\quad 2n^2 + 3 = 2 \times 1 + 3 = 5$ Prime

$n = 2{:}\quad 2n^2 + 3 = 2 \times 4 + 3 = 11$ Prime

$n = 3{:}\quad 2n^2 + 3 = 2 \times 9 + 3 = 21 = 3 \times 7$ Not prime

Now try this

1 Tim is x years old.
Kelly is 5 years younger than Tim.
Ben is twice as old as Tim.
The total of their ages is 31 years.
Form an equation in x and use it to work out Tim's age. *(4 marks)*

2 Jill says that the expression $n^2 + 4n - 3$, where n is a **whole number**, is never a multiple of 7.

Give a counter example to show that Jill is wrong. *(2 marks)*

C D E F G

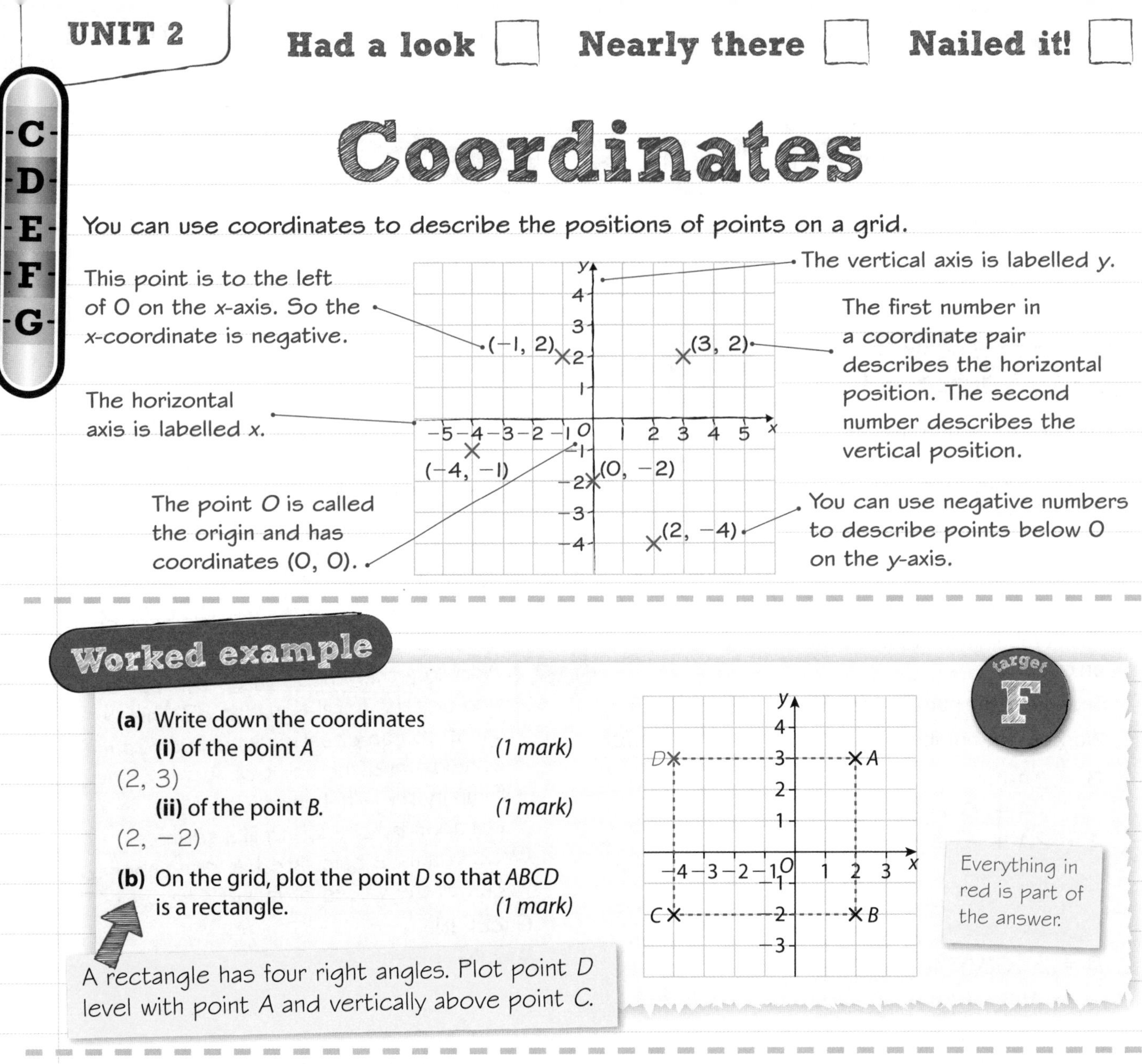

Coordinates

You can use coordinates to describe the positions of points on a grid.

Worked example

Target F

(a) Write down the coordinates

(i) of the point *A* *(1 mark)*

(2, 3)

(ii) of the point *B*. *(1 mark)*

(2, −2)

(b) On the grid, plot the point *D* so that *ABCD* is a rectangle. *(1 mark)*

Everything in red is part of the answer.

A rectangle has four right angles. Plot point *D* level with point *A* and vertically above point *C*.

Midpoints

A LINE SEGMENT is a short section of a straight line. The midpoint of a line segment is exactly halfway along the line. You can find the midpoint if you know the coordinates of the ends.

To find the midpoint, add the *x*-coordinates and divide by 2 and add the *y*-coordinates and divide by 2.

$$\text{Midpoint} = \left(\frac{x_1 + x_2}{2}, \frac{y_1 + y_2}{2}\right)$$

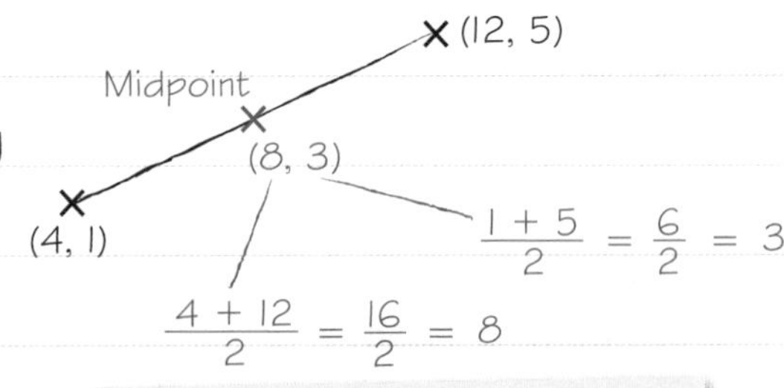

You need to be confident in finding midpoints if you're going for a grade D or C.

Now try this

Target G-D

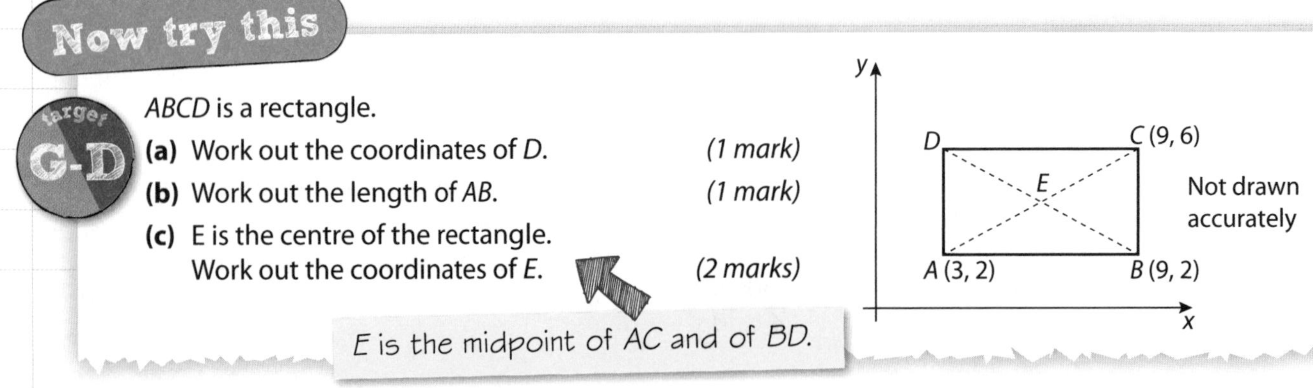

ABCD is a rectangle.

(a) Work out the coordinates of *D*. *(1 mark)*

(b) Work out the length of *AB*. *(1 mark)*

(c) E is the centre of the rectangle. Work out the coordinates of *E*. *(2 marks)*

E is the midpoint of *AC* and of *BD*.

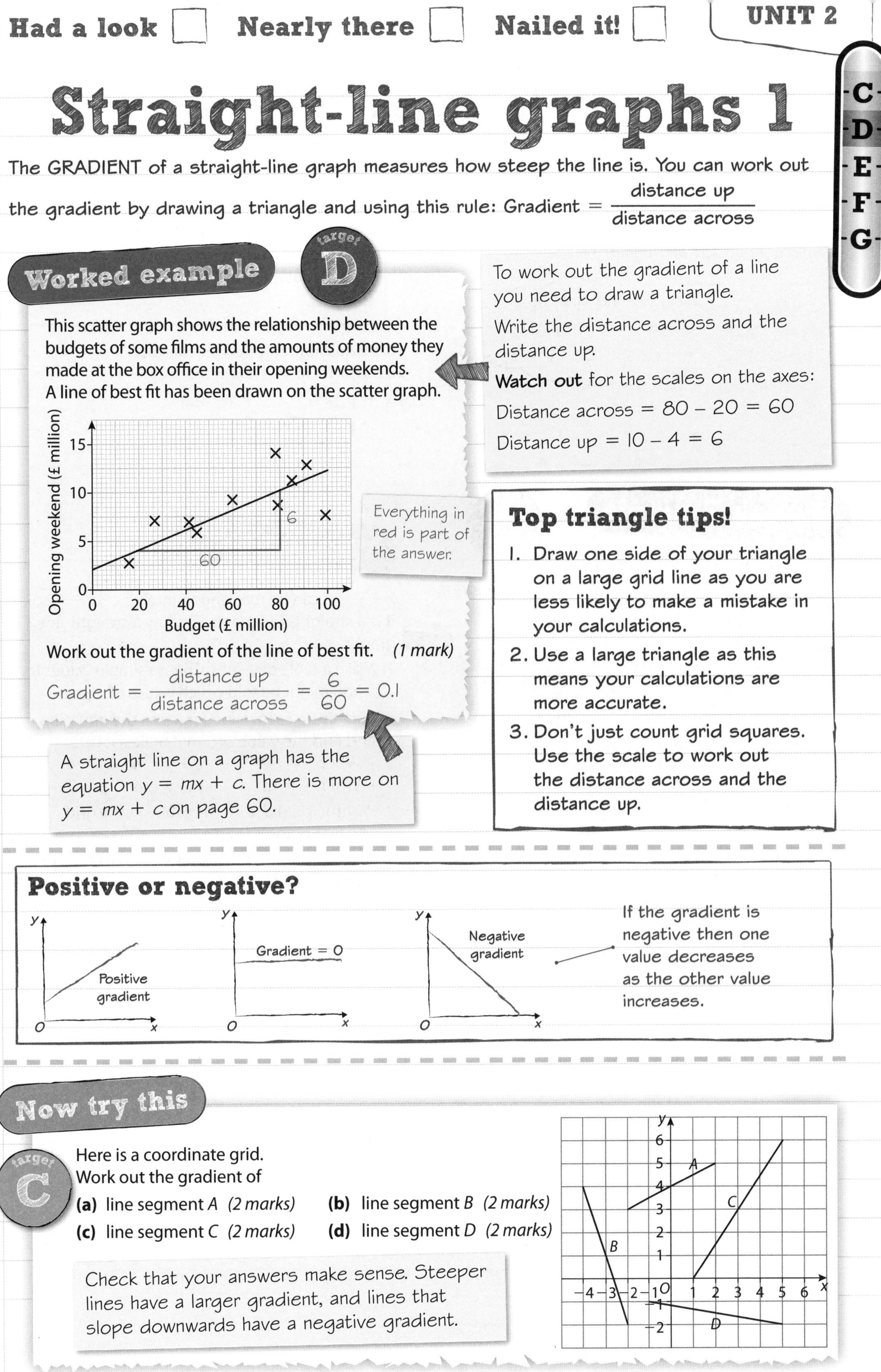

Straight-line graphs 1

C D E F G

The GRADIENT of a straight-line graph measures how steep the line is. You can work out the gradient by drawing a triangle and using this rule: Gradient $= \frac{\text{distance up}}{\text{distance across}}$

Worked example (target D)

This scatter graph shows the relationship between the budgets of some films and the amounts of money they made at the box office in their opening weekends.
A line of best fit has been drawn on the scatter graph.

Work out the gradient of the line of best fit. *(1 mark)*

Gradient $= \frac{\text{distance up}}{\text{distance across}} = \frac{6}{60} = 0.1$

To work out the gradient of a line you need to draw a triangle.
Write the distance across and the distance up.
Watch out for the scales on the axes:
Distance across = 80 − 20 = 60
Distance up = 10 − 4 = 6

Everything in red is part of the answer.

A straight line on a graph has the equation $y = mx + c$. There is more on $y = mx + c$ on page 60.

Top triangle tips!

1. Draw one side of your triangle on a large grid line as you are less likely to make a mistake in your calculations.
2. Use a large triangle as this means your calculations are more accurate.
3. Don't just count grid squares. Use the scale to work out the distance across and the distance up.

Positive or negative?

If the gradient is negative then one value decreases as the other value increases.

Now try this (target C)

Here is a coordinate grid.
Work out the gradient of

(a) line segment *A* *(2 marks)* **(b)** line segment *B* *(2 marks)*
(c) line segment *C* *(2 marks)* **(d)** line segment *D* *(2 marks)*

Check that your answers make sense. Steeper lines have a larger gradient, and lines that slope downwards have a negative gradient.

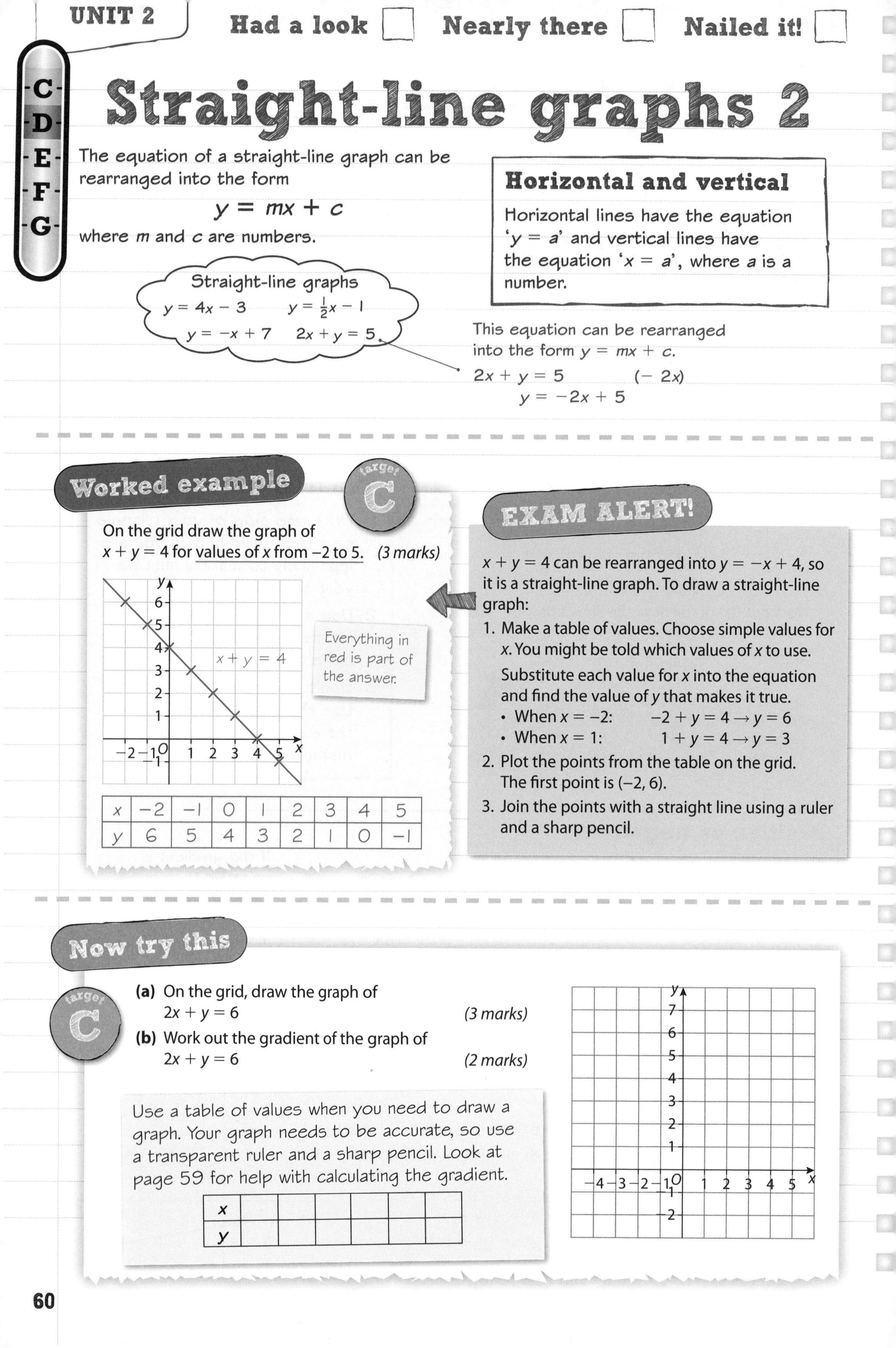

C D E F G

Straight-line graphs 2

The equation of a straight-line graph can be rearranged into the form

$$y = mx + c$$

where m and c are numbers.

This equation can be rearranged into the form $y = mx + c$.

$2x + y = 5$ $(-2x)$

$y = -2x + 5$

Horizontal and vertical

Horizontal lines have the equation '$y = a$' and vertical lines have the equation '$x = a$', where a is a number.

Worked example

target C

On the grid draw the graph of $x + y = 4$ for values of x from -2 to 5. *(3 marks)*

Everything in red is part of the answer.

x	−2	−1	0	1	2	3	4	5
y	6	5	4	3	2	1	0	−1

EXAM ALERT!

$x + y = 4$ can be rearranged into $y = -x + 4$, so it is a straight-line graph. To draw a straight-line graph:

1. Make a table of values. Choose simple values for x. You might be told which values of x to use.
   Substitute each value for x into the equation and find the value of y that makes it true.
   - When $x = -2$: $-2 + y = 4 \rightarrow y = 6$
   - When $x = 1$: $1 + y = 4 \rightarrow y = 3$
2. Plot the points from the table on the grid. The first point is $(-2, 6)$.
3. Join the points with a straight line using a ruler and a sharp pencil.

Now try this

target C

(a) On the grid, draw the graph of $2x + y = 6$ *(3 marks)*

(b) Work out the gradient of the graph of $2x + y = 6$ *(2 marks)*

Use a table of values when you need to draw a graph. Your graph needs to be accurate, so use a transparent ruler and a sharp pencil. Look at page 59 for help with calculating the gradient.

x						
y						

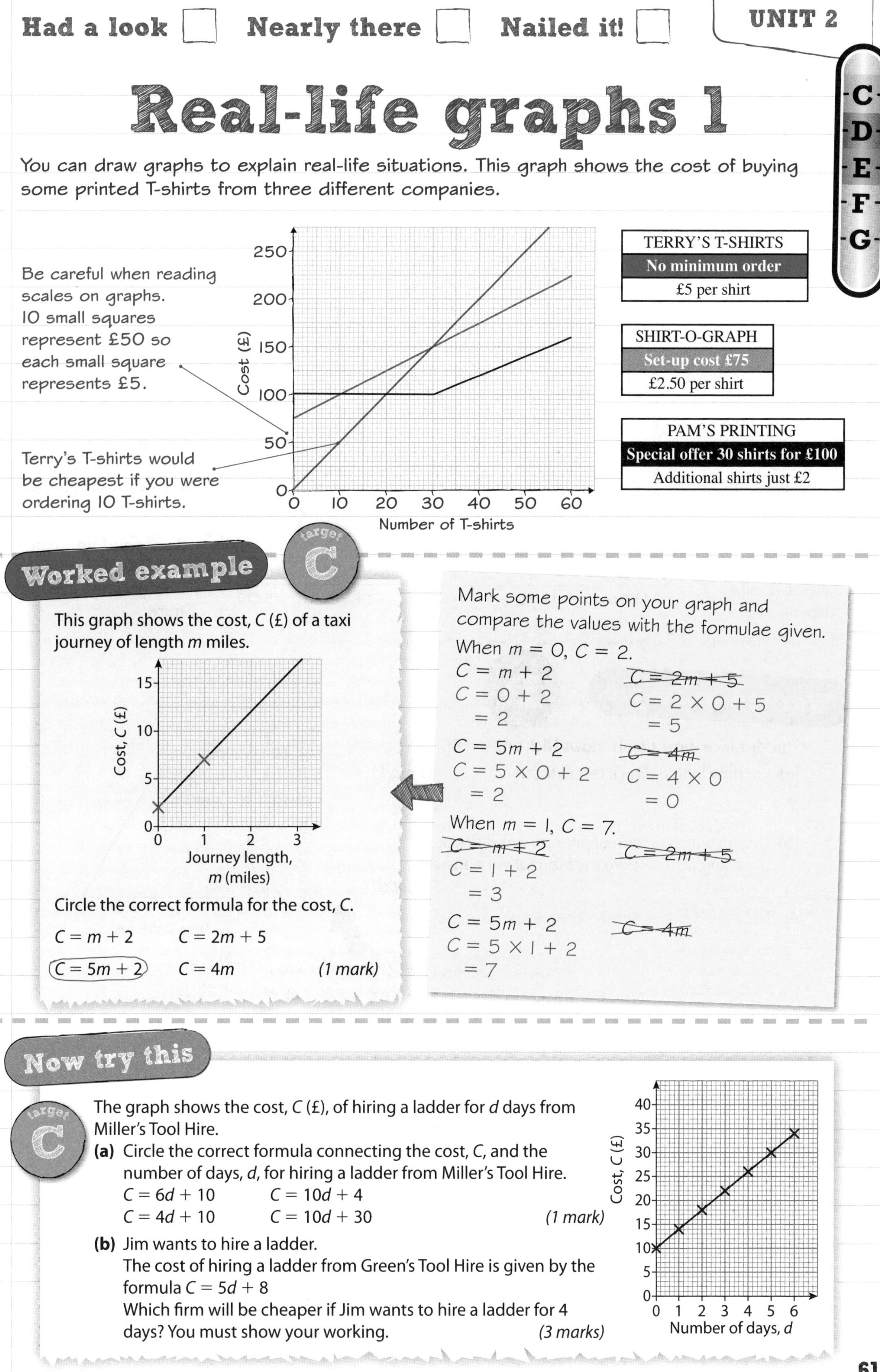

C D E F G

Real-life graphs 1

You can draw graphs to explain real-life situations. This graph shows the cost of buying some printed T-shirts from three different companies.

TERRY'S T-SHIRTS
No minimum order
£5 per shirt

SHIRT-O-GRAPH
Set-up cost £75
£2.50 per shirt

PAM'S PRINTING
Special offer 30 shirts for £100
Additional shirts just £2

Be careful when reading scales on graphs. 10 small squares represent £50 so each small square represents £5.

Terry's T-shirts would be cheapest if you were ordering 10 T-shirts.

Worked example

target C

This graph shows the cost, C (£) of a taxi journey of length m miles.

Circle the correct formula for the cost, C.

$C = m + 2$ $C = 2m + 5$

($C = 5m + 2$) $C = 4m$ *(1 mark)*

Mark some points on your graph and compare the values with the formulae given.

When $m = 0$, $C = 2$.

$C = m + 2$
$C = 0 + 2$
$= 2$

~~$C = 2m + 5$~~
$C = 2 \times 0 + 5$
$= 5$

$C = 5m + 2$
$C = 5 \times 0 + 2$
$= 2$

~~$C = 4m$~~
$C = 4 \times 0$
$= 0$

When $m = 1$, $C = 7$.

~~$C = m + 2$~~
$C = 1 + 2$
$= 3$

~~$C = 2m + 5$~~

$C = 5m + 2$
$C = 5 \times 1 + 2$
$= 7$

~~$C = 4m$~~

Now try this

target C

The graph shows the cost, C (£), of hiring a ladder for d days from Miller's Tool Hire.

(a) Circle the correct formula connecting the cost, C, and the number of days, d, for hiring a ladder from Miller's Tool Hire.

$C = 6d + 10$ $C = 10d + 4$

$C = 4d + 10$ $C = 10d + 30$ *(1 mark)*

(b) Jim wants to hire a ladder.
The cost of hiring a ladder from Green's Tool Hire is given by the formula $C = 5d + 8$
Which firm will be cheaper if Jim wants to hire a ladder for 4 days? You must show your working. *(3 marks)*

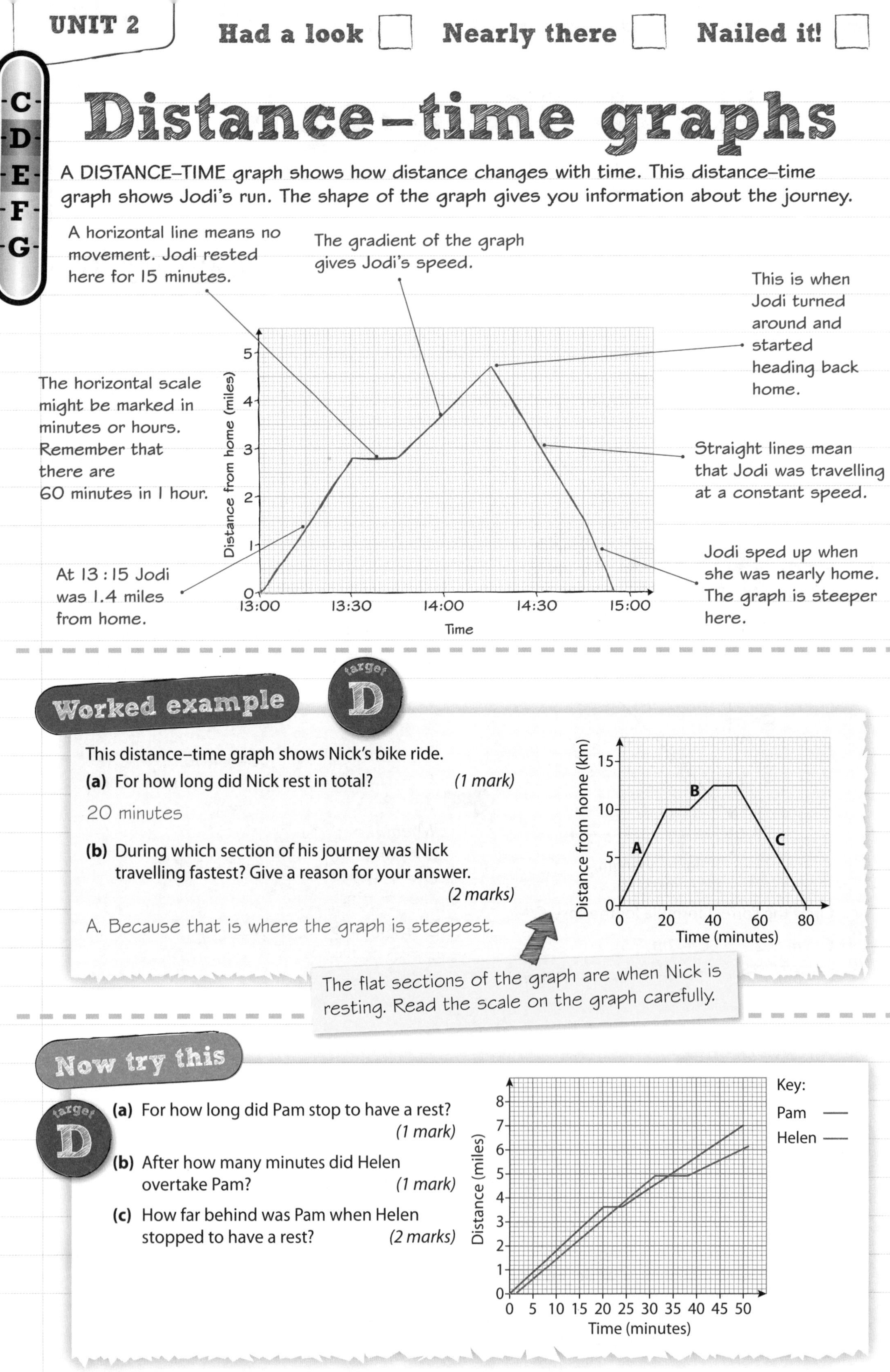

C D E F G

Distance–time graphs

A DISTANCE–TIME graph shows how distance changes with time. This distance–time graph shows Jodi's run. The shape of the graph gives you information about the journey.

Worked example

target D

This distance–time graph shows Nick's bike ride.

(a) For how long did Nick rest in total? *(1 mark)*

20 minutes

(b) During which section of his journey was Nick travelling fastest? Give a reason for your answer. *(2 marks)*

A. Because that is where the graph is steepest.

The flat sections of the graph are when Nick is resting. Read the scale on the graph carefully.

Now try this

target D

(a) For how long did Pam stop to have a rest? *(1 mark)*

(b) After how many minutes did Helen overtake Pam? *(1 mark)*

(c) How far behind was Pam when Helen stopped to have a rest? *(2 marks)*

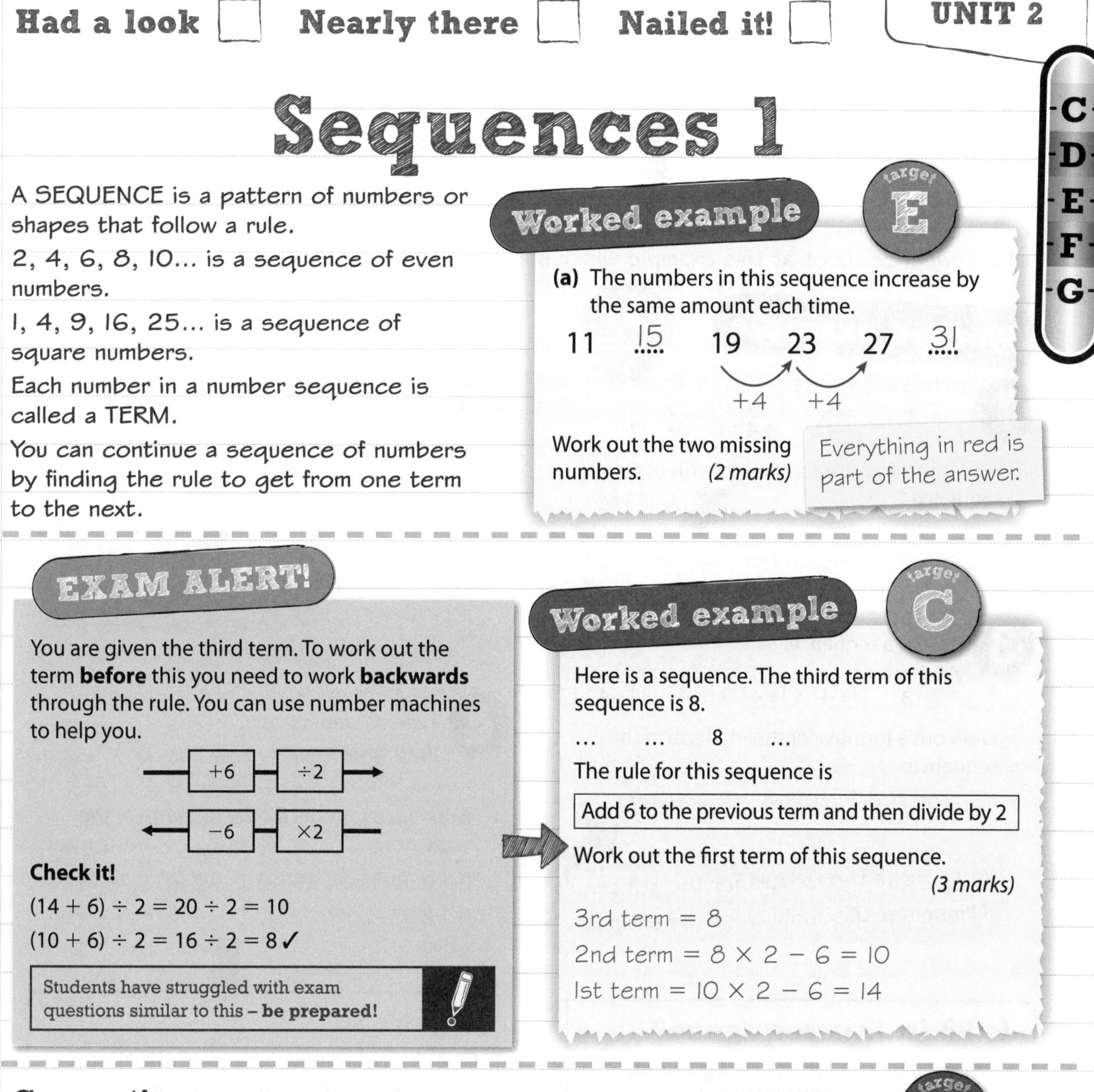

C D E F G

Sequences 1

A SEQUENCE is a pattern of numbers or shapes that follow a rule.

2, 4, 6, 8, 10... is a sequence of even numbers.

1, 4, 9, 16, 25... is a sequence of square numbers.

Each number in a number sequence is called a TERM.

You can continue a sequence of numbers by finding the rule to get from one term to the next.

Worked example (target E)

(a) The numbers in this sequence increase by the same amount each time.

11 15 19 23 27 31

+4 +4

Work out the two missing numbers. *(2 marks)*

Everything in red is part of the answer.

EXAM ALERT!

You are given the third term. To work out the term **before** this you need to work **backwards** through the rule. You can use number machines to help you.

→ +6 → ÷2 →

← −6 ← ×2 ←

Check it!

$(14 + 6) \div 2 = 20 \div 2 = 10$

$(10 + 6) \div 2 = 16 \div 2 = 8$ ✓

Students have struggled with exam questions similar to this – **be prepared!**

Worked example (target C)

Here is a sequence. The third term of this sequence is 8.

… … 8 …

The rule for this sequence is

Add 6 to the previous term and then divide by 2

Work out the first term of this sequence. *(3 marks)*

3rd term = 8

2nd term = $8 \times 2 - 6 = 10$

1st term = $10 \times 2 - 6 = 14$

Generating sequences

You can work out the terms of a sequence by substituting the term number into the *n*th term. Here are some examples:

*n*th term	$9 - 2n$	$n^2 + 10$
1st term	$9 - 2 \times 1 = 7$	$1^2 + 10 = 11$
2nd term	$9 - 2 \times 2 = 5$	$2^2 + 10 = 14$
↓	↓	↓
8th term	$9 - 2 \times 8 = -7$	$8^2 + 10 = 74$

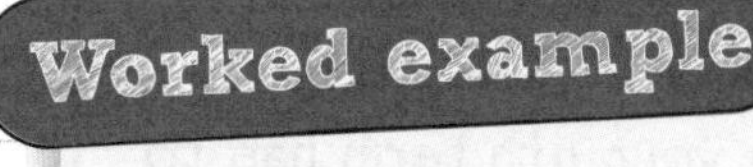

Worked example (target D)

The *n*th term of a sequence is $50 - n^2$.

Work out the first term of the sequence that is negative. *(2 marks)*

7th term = $50 - 7^2 = 1$

8th term = $50 - 8^2 = -14$

The 8th term is the first negative term.

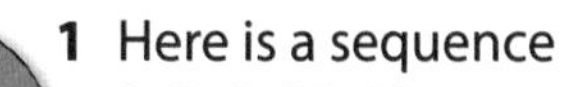

Now try this

(target E)

1 Here is a sequence

3, 5, 9, 15, 23, …, …

(a) Describe the term-to-term rule for this sequence. *(1 mark)*

(b) Write down the next **two** terms in the sequence. *(2 marks)*

(target D)

2 The *n*th term of a sequence is $4n + 27$

(a) Work out the first **three** terms. *(2 marks)*

(b) Work out which term of the sequence is the first one greater than 100. *(2 marks)*

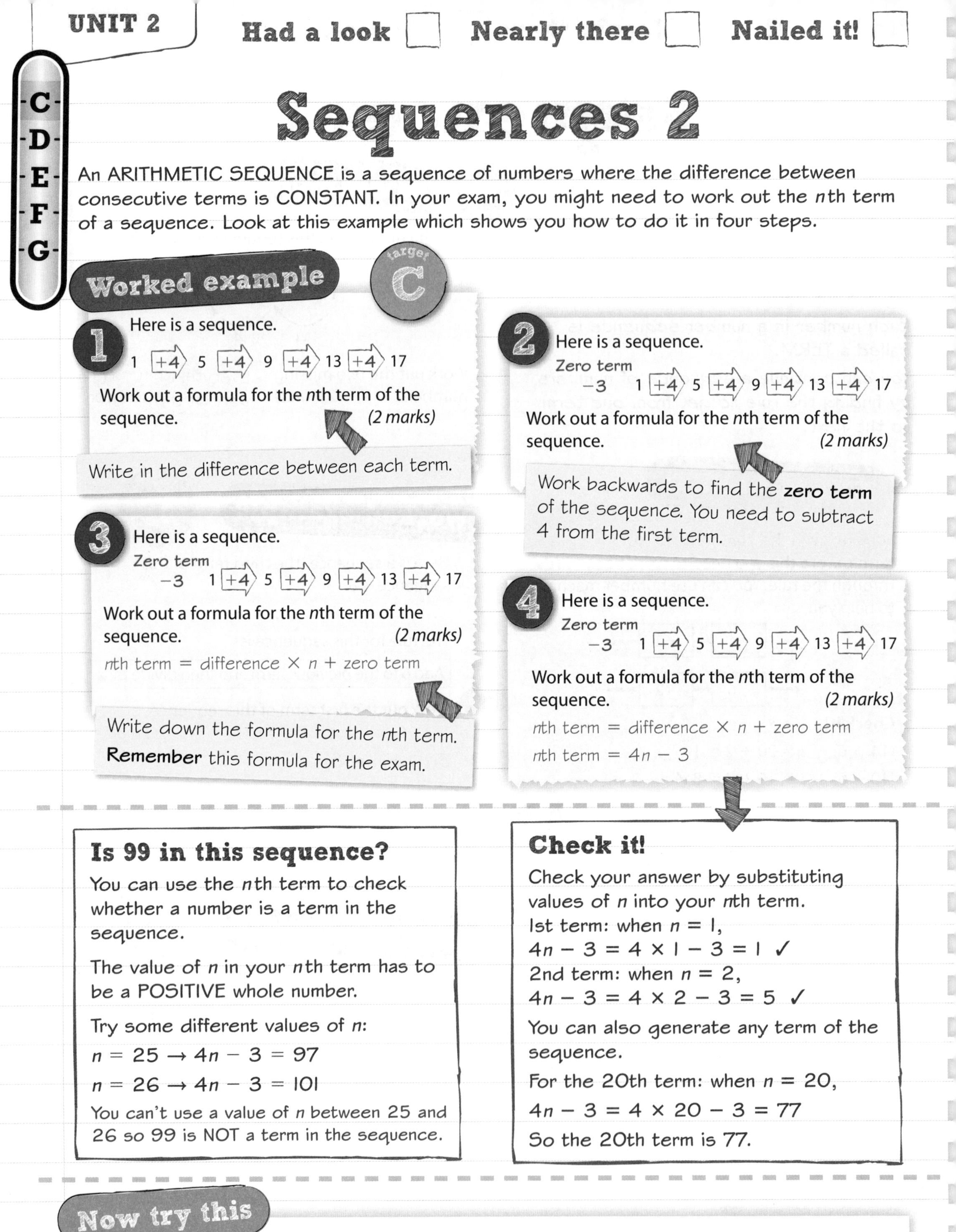

C D E F G

Sequences 2

An ARITHMETIC SEQUENCE is a sequence of numbers where the difference between consecutive terms is CONSTANT. In your exam, you might need to work out the nth term of a sequence. Look at this example which shows you how to do it in four steps.

Worked example

target C

1 Here is a sequence.

1 (+4) 5 (+4) 9 (+4) 13 (+4) 17

Work out a formula for the nth term of the sequence. *(2 marks)*

Write in the difference between each term.

2 Here is a sequence.

Zero term −3 1 (+4) 5 (+4) 9 (+4) 13 (+4) 17

Work out a formula for the nth term of the sequence. *(2 marks)*

Work backwards to find the **zero term** of the sequence. You need to subtract 4 from the first term.

3 Here is a sequence.

Zero term −3 1 (+4) 5 (+4) 9 (+4) 13 (+4) 17

Work out a formula for the nth term of the sequence. *(2 marks)*

nth term = difference × n + zero term

Write down the formula for the nth term. **Remember** this formula for the exam.

4 Here is a sequence.

Zero term −3 1 (+4) 5 (+4) 9 (+4) 13 (+4) 17

Work out a formula for the nth term of the sequence. *(2 marks)*

nth term = difference × n + zero term

nth term = $4n - 3$

Is 99 in this sequence?

You can use the nth term to check whether a number is a term in the sequence.

The value of n in your nth term has to be a POSITIVE whole number.

Try some different values of n:

$n = 25 \rightarrow 4n - 3 = 97$

$n = 26 \rightarrow 4n - 3 = 101$

You can't use a value of n between 25 and 26 so 99 is NOT a term in the sequence.

Check it!

Check your answer by substituting values of n into your nth term.

1st term: when $n = 1$,
$4n - 3 = 4 \times 1 - 3 = 1$ ✓

2nd term: when $n = 2$,
$4n - 3 = 4 \times 2 - 3 = 5$ ✓

You can also generate any term of the sequence.

For the 20th term: when $n = 20$,

$4n - 3 = 4 \times 20 - 3 = 77$

So the 20th term is 77.

Now try this

target C

Here is a sequence: 2 5 8 11 14 ……

(a) Work out an expression for the nth term of the sequence. *(2 marks)*

(b) Work out the 29th term in the sequence. *(2 marks)*

(c) How many terms of this sequence are **less than** 200? *(2 marks)*

(d) Is 156 a term in this sequence? *(2 marks)*

Problem-solving practice 1

About half of the questions on your exam will need problem-solving skills.
These skills are sometimes called AO2 and AO3.
Practise using the questions on the next two pages.
For these questions you might need to:

- choose which mathematical technique or skill to use
- apply a technique in a new context
- plan your strategy to solve a longer problem
- show your working clearly and give reasons for your answers.

AO2 AO3

*Here are five numbers

7 12 23 30 75

Make a fraction with a value between 2 and 3.

Use one of the numbers for the numerator and one of the numbers for the denominator. *(2 marks)*

Mixed numbers p. 39

You need to make an improper fraction. If you convert your improper fraction into a mixed number you can show that it is between 2 and 3.

If a question has a * next to it then one mark is awarded for **quality of written communication**. In this question you need to write down the correct fraction **and** show that its value is between 2 and 3.

Here are some patterns made from dots.

Pattern number 1 Pattern number 2 Pattern number 3

(a) How many dots are needed for Pattern number 5? *(1 mark)*

(b) Vidya says she can make a pattern out of 21 dots.

Is Vidya correct? You must give a reason for your answer. *(2 marks)*

Sequences 1 p. 63

For part **(a)** you should draw a table showing the number of dots in each pattern. Carry on the number sequence until you get to Pattern number 5. For part **(b)** you should look to see if 21 could be a number in the sequence. If it can't, make sure you give a reason why not.

TOP TIP

Check your number sequence by drawing the next pattern and counting the dots.

Problem-solving practice 2

Suresh wants to buy a new pair of trainers.

There are three shops that sell the trainers he wants.

Sportcentre Trainers	Footwear First Trainers	Action Sport Trainers
£10 plus 10 payments of £3.50	$\frac{1}{4}$ off usual price of £80	£40 plus VAT at 20%

Which shop is selling the trainers the cheapest? *(5 marks)*

Fractions, decimals and percentages p. 40

There are lots of steps in this question so make sure you keep track of your working. You need to calculate the price at each shop, then write down which shop is cheapest.

Divide the answer space into three columns. Then the examiner can see which shop each bit of working is for.

Jamie, Amir and Helen are comparing their ages.

Jamie is six years older than Amir.

Helen is three times older than Amir.

The sum of all three of their ages is 41 years.

Let x be Amir's age.

Set up and solve an equation to find Amir's age. *(3 marks)*

Using algebra p. 57

Write expressions for Jamie's age and Helen's age in terms of x. Add together all three expressions (including x for Amir's age) and set this new expression equal to 41. You can solve this equation to find x.

If the question tells you to use an equation you will not get full marks if you use trial and improvement.

On the grid, draw the graph of $x + y = 5$ for values of x from -3 to 4

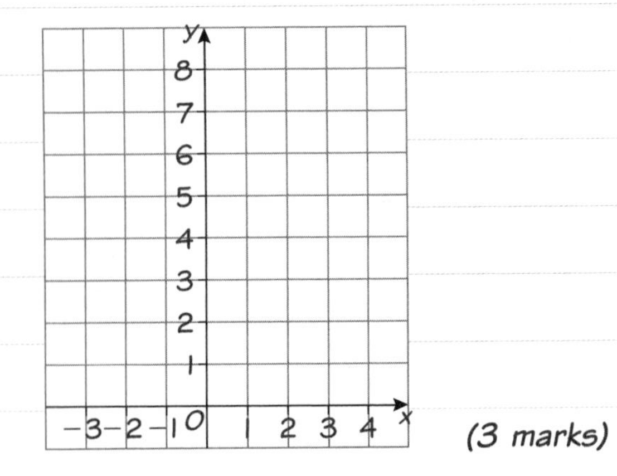

(3 marks)

Straight-line graphs 2 p. 60

Start by drawing a table of values.

x	−3	−2	−1	0	1	2	3	4
y							2	

Substitute each value for x into the equation and find the value of y that makes it true.

TOP TIP

It's easier to start with the positive values of x. Look to see if the values of y in the table follow a pattern.

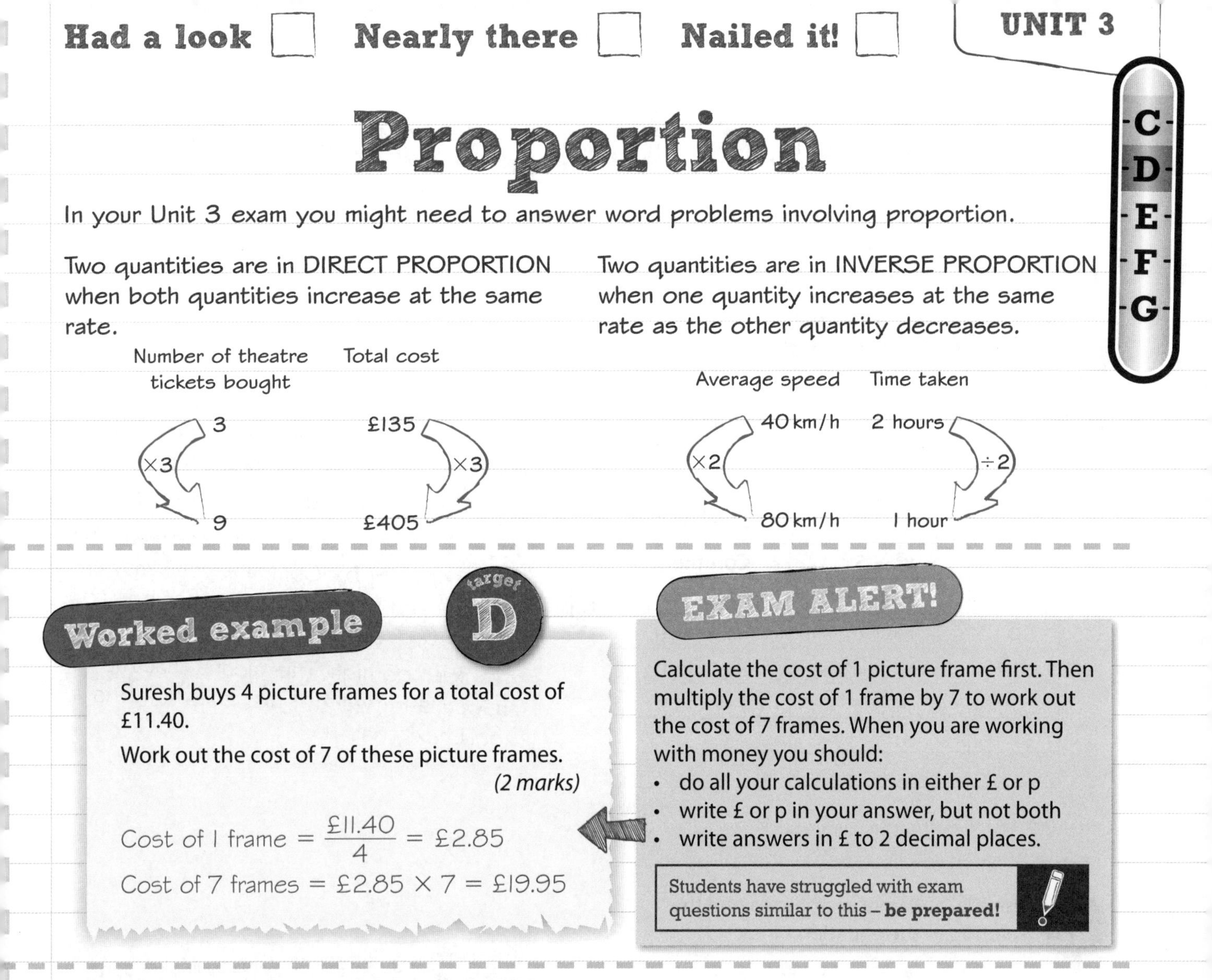

Proportion

In your Unit 3 exam you might need to answer word problems involving proportion.

Two quantities are in DIRECT PROPORTION when both quantities increase at the same rate.

Two quantities are in INVERSE PROPORTION when one quantity increases at the same rate as the other quantity decreases.

Worked example

target D

Suresh buys 4 picture frames for a total cost of £11.40.

Work out the cost of 7 of these picture frames. *(2 marks)*

Cost of 1 frame $= \frac{£11.40}{4} = £2.85$

Cost of 7 frames $= £2.85 \times 7 = £19.95$

EXAM ALERT!

Calculate the cost of 1 picture frame first. Then multiply the cost of 1 frame by 7 to work out the cost of 7 frames. When you are working with money you should:

- do all your calculations in either £ or p
- write £ or p in your answer, but not both
- write answers in £ to 2 decimal places.

Students have struggled with exam questions similar to this – **be prepared!**

Divide or multiply?

6 people can build a wall in 4 days.
How long would it take 8 people to build the same wall?

Inverse proportion problems often involve time. The more people working on a task, the quicker it will be finished.

You can solve this problem by working out how long it would take 1 person to build the wall. Use common sense to decide whether to divide or multiply.

$6 \times 4 = 24$ so 1 person could build the wall in 24 days.

You multiply because it would take 1 person more time to build the wall.

$24 \div 8 = 3$ so 8 people could build the wall in 3 days.

You divide because it would take 8 people less time to build the wall.

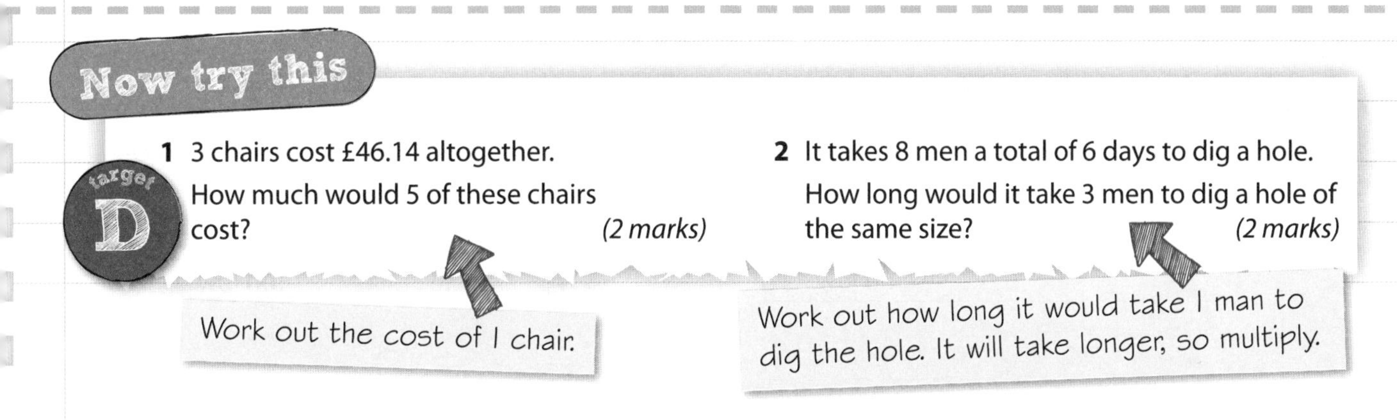

Now try this

target D

1 3 chairs cost £46.14 altogether.
How much would 5 of these chairs cost? *(2 marks)*

2 It takes 8 men a total of 6 days to dig a hole.
How long would it take 3 men to dig a hole of the same size? *(2 marks)*

Work out the cost of 1 chair.

Work out how long it would take 1 man to dig the hole. It will take longer, so multiply.

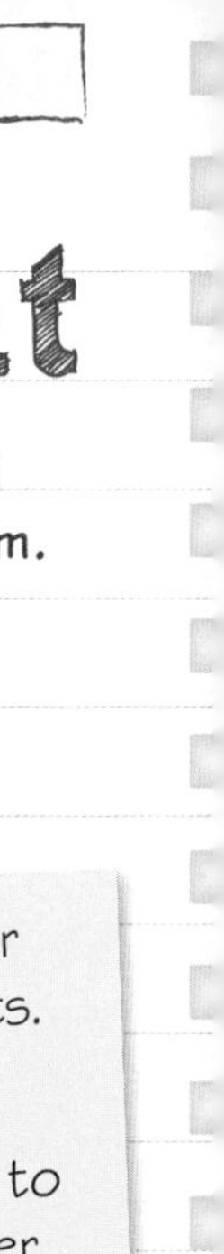

Trial and improvement

Some equations can't be solved exactly. You need to use trial and improvement to find an approximate solution. You will be told when to use trial and improvement in your exam. Look at this worked example which shows you how to do it in two steps.

Worked example

target C

1 Use trial and improvement to find the solution to the equation

$$x^3 - 5x = 60$$

Give your answer correct to 1 decimal place.

x	$x^3 - 5x$	Comment
4	44	Too low
5	100	Too high
4.5	68.625	Too high

(4 marks)

You will usually be given one trial in your exam, and a table to record your results. $x = 4$ is too low, so try $x = 5$.

You can use the x^3 on your calculator to work out $x^3 - 5x$. Compare your answer with 60 and write down whether it is too high or too low.

$x = 5$ is too high. This means that the answer is between 4 and 5. $x = 4.5$ is a good next trial.

2 Use trial and improvement to find the solution to the equation

$$x^3 - 5x = 60$$

Give your answer correct to 1 decimal place.

x	$x^3 - 5x$	Comment
4	44	Too low
5	100	Too high
4.5	68.625	Too high
4.3	58.007	Too low
4.4	63.184	Too high
4.35	60.562...	Too high

(4 marks)

$x = 4.3$ (to 1 decimal place)

Keep trying different values. Make sure you write down the results of every trial.

$x = 4.3$ is too low and $x = 4.4$ is too high, so you know the answer is between 4.3 and 4.4. But you don't know which value is closer.

Try $x = 4.35$

$x = 4.35$ is too high, so you know the answer is between $x = 4.3$ and $x = 4.35$ This means it is closer to $x = 4.3$

Write down the answer correct to 1 decimal place.

Now try this

1 Use trial and improvement to find the solution to $2x^2 + 5x = 29$.
Give your answer correct to 1 decimal place. *(4 marks)*

x	$2x^2 + 5x$	Comment
2	18	Too low

Remember $2x^2$ means $2 \times x^2$

2 Use trial and improvement to find the solution to $x(x + 3)(x - 1) = 54$.
Give your answer correct to 1 decimal place. *(4 marks)*

x	$x(x + 3)(x - 1)$	Comment
3	36	Too low

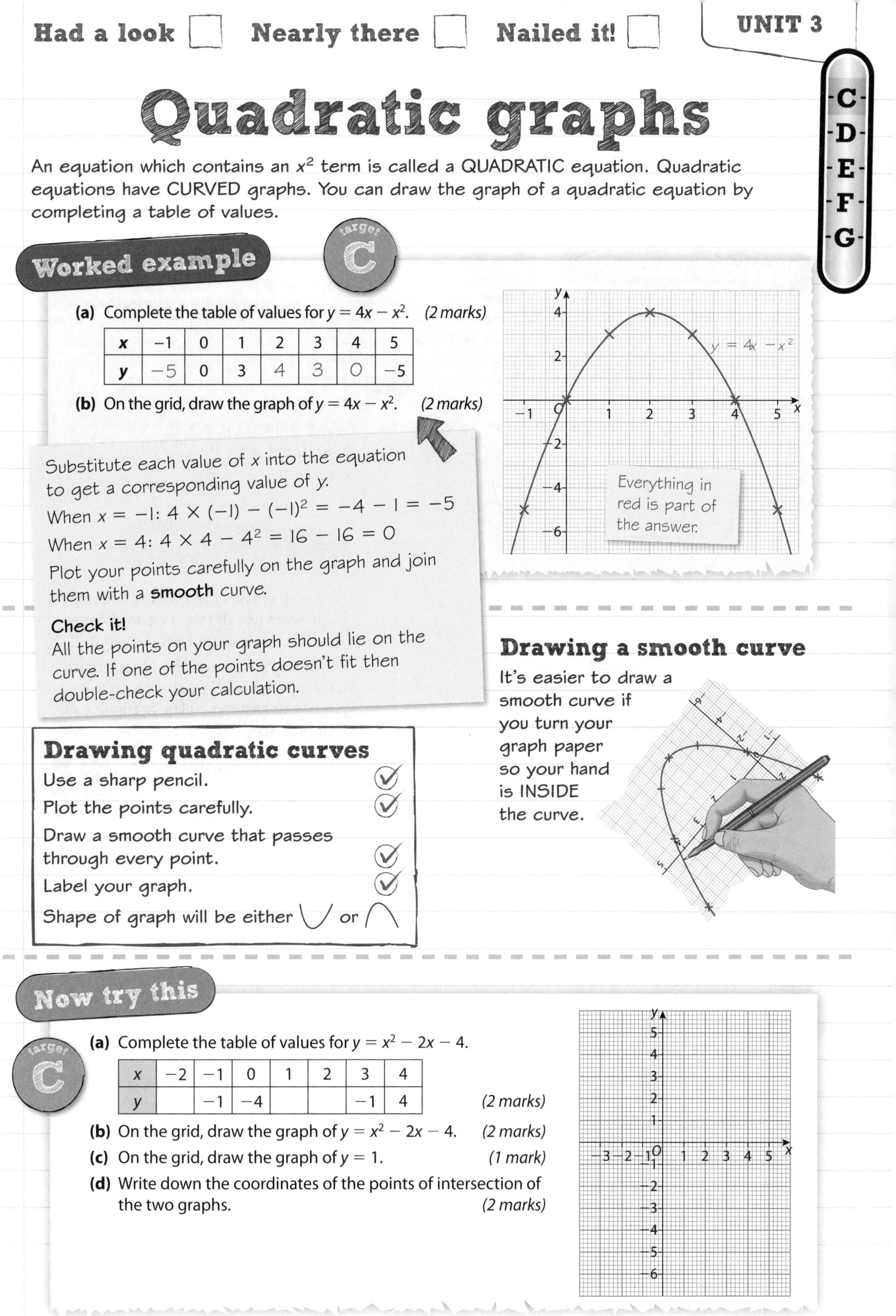

C D E F G

Quadratic graphs

An equation which contains an x^2 term is called a QUADRATIC equation. Quadratic equations have CURVED graphs. You can draw the graph of a quadratic equation by completing a table of values.

Worked example

target C

(a) Complete the table of values for $y = 4x - x^2$. *(2 marks)*

x	−1	0	1	2	3	4	5
y	−5	0	3	4	3	0	−5

(b) On the grid, draw the graph of $y = 4x - x^2$. *(2 marks)*

Substitute each value of x into the equation to get a corresponding value of y.

When $x = -1$: $4 \times (-1) - (-1)^2 = -4 - 1 = -5$

When $x = 4$: $4 \times 4 - 4^2 = 16 - 16 = 0$

Plot your points carefully on the graph and join them with a **smooth** curve.

Check it!
All the points on your graph should lie on the curve. If one of the points doesn't fit then double-check your calculation.

Drawing quadratic curves

- Use a sharp pencil. ✓
- Plot the points carefully. ✓
- Draw a smooth curve that passes through every point. ✓
- Label your graph. ✓
- Shape of graph will be either ∪ or ∩

Drawing a smooth curve

It's easier to draw a smooth curve if you turn your graph paper so your hand is INSIDE the curve.

Now try this

target C

(a) Complete the table of values for $y = x^2 - 2x - 4$.

x	−2	−1	0	1	2	3	4
y		−1	−4			−1	4

(2 marks)

(b) On the grid, draw the graph of $y = x^2 - 2x - 4$. *(2 marks)*

(c) On the grid, draw the graph of $y = 1$. *(1 mark)*

(d) Write down the coordinates of the points of intersection of the two graphs. *(2 marks)*

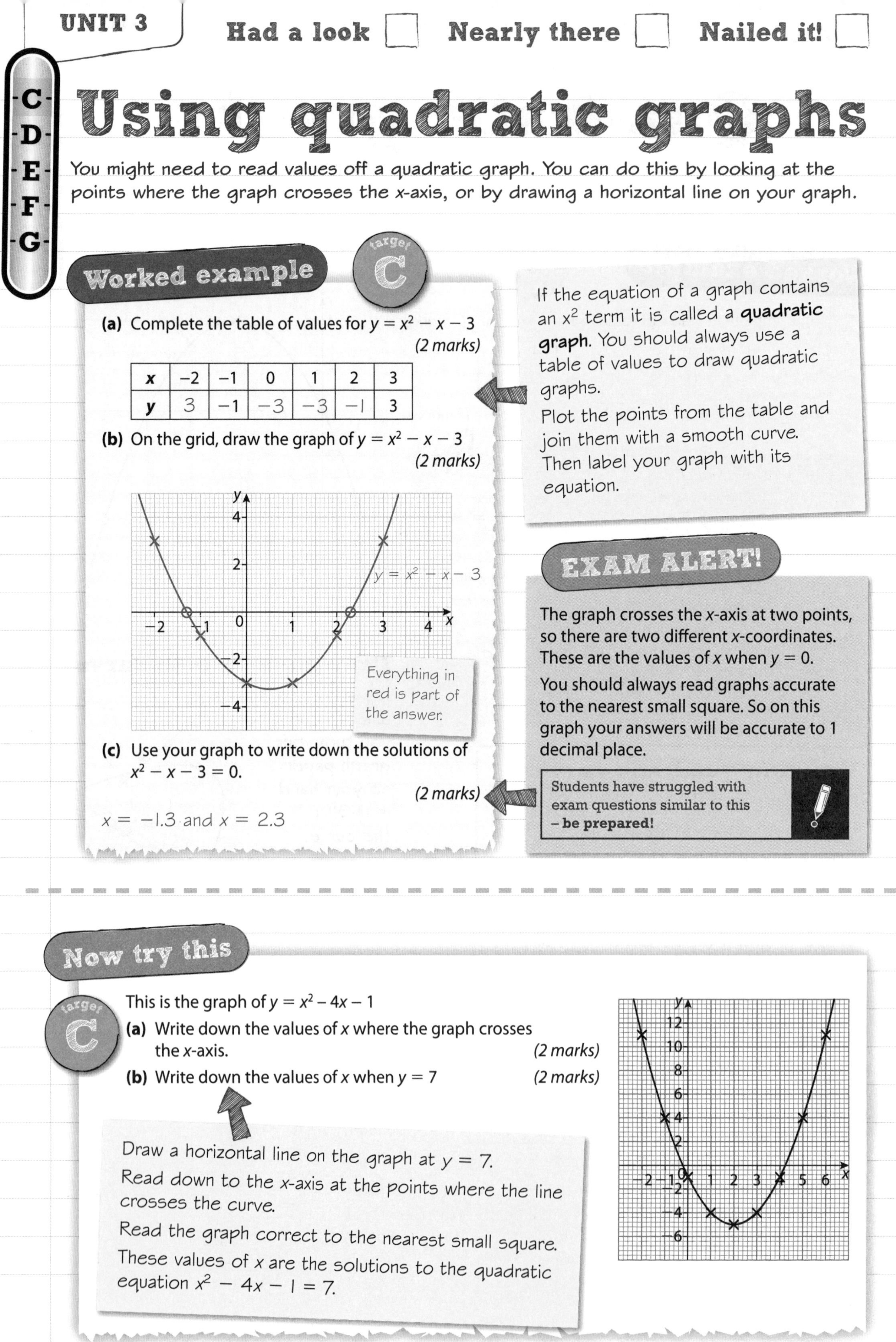

C D E F G

Using quadratic graphs

You might need to read values off a quadratic graph. You can do this by looking at the points where the graph crosses the x-axis, or by drawing a horizontal line on your graph.

Worked example (target C)

(a) Complete the table of values for $y = x^2 - x - 3$ *(2 marks)*

x	−2	−1	0	1	2	3
y	3	−1	−3	−3	−1	3

(b) On the grid, draw the graph of $y = x^2 - x - 3$ *(2 marks)*

(c) Use your graph to write down the solutions of $x^2 - x - 3 = 0$. *(2 marks)*

$x = -1.3$ and $x = 2.3$

If the equation of a graph contains an x^2 term it is called a **quadratic graph**. You should always use a table of values to draw quadratic graphs.

Plot the points from the table and join them with a smooth curve. Then label your graph with its equation.

EXAM ALERT!

The graph crosses the x-axis at two points, so there are two different x-coordinates. These are the values of x when $y = 0$.

You should always read graphs accurate to the nearest small square. So on this graph your answers will be accurate to 1 decimal place.

Students have struggled with exam questions similar to this **– be prepared!**

Now try this (target C)

This is the graph of $y = x^2 - 4x - 1$

(a) Write down the values of x where the graph crosses the x-axis. *(2 marks)*

(b) Write down the values of x when $y = 7$ *(2 marks)*

Draw a horizontal line on the graph at $y = 7$.

Read down to the x-axis at the points where the line crosses the curve.

Read the graph correct to the nearest small square.

These values of x are the solutions to the quadratic equation $x^2 - 4x - 1 = 7$.

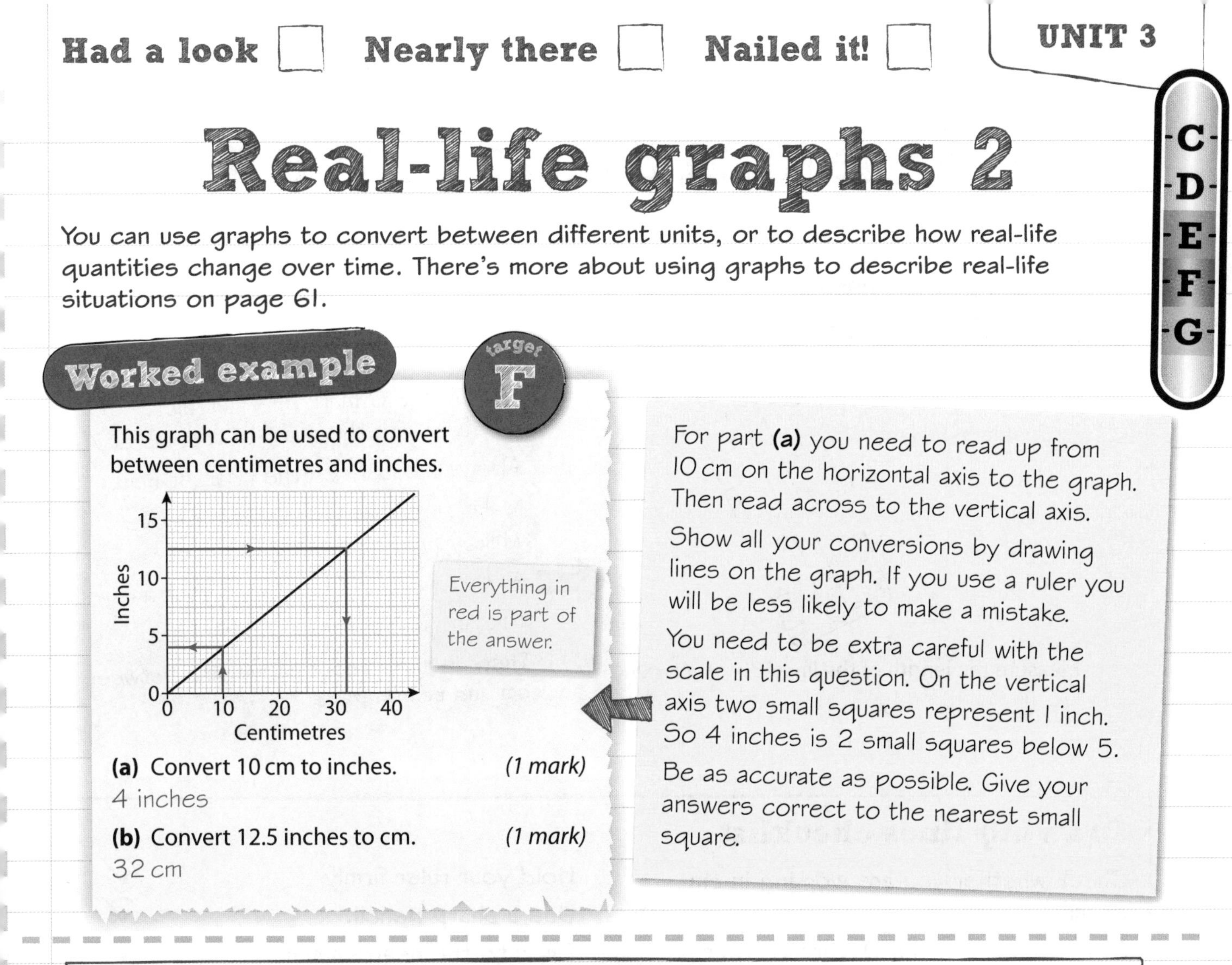

Real-life graphs 2

You can use graphs to convert between different units, or to describe how real-life quantities change over time. There's more about using graphs to describe real-life situations on page 61.

Worked example

target F

This graph can be used to convert between centimetres and inches.

Everything in red is part of the answer.

(a) Convert 10 cm to inches. *(1 mark)*

4 inches

(b) Convert 12.5 inches to cm. *(1 mark)*

32 cm

For part **(a)** you need to read up from 10 cm on the horizontal axis to the graph. Then read across to the vertical axis.

Show all your conversions by drawing lines on the graph. If you use a ruler you will be less likely to make a mistake.

You need to be extra careful with the scale in this question. On the vertical axis two small squares represent 1 inch. So 4 inches is 2 small squares below 5.

Be as accurate as possible. Give your answers correct to the nearest small square.

Quantities which change with time

These garden ponds are being filled with water at a constant rate.
The graphs below show how the depth of water in each pond changes with time.

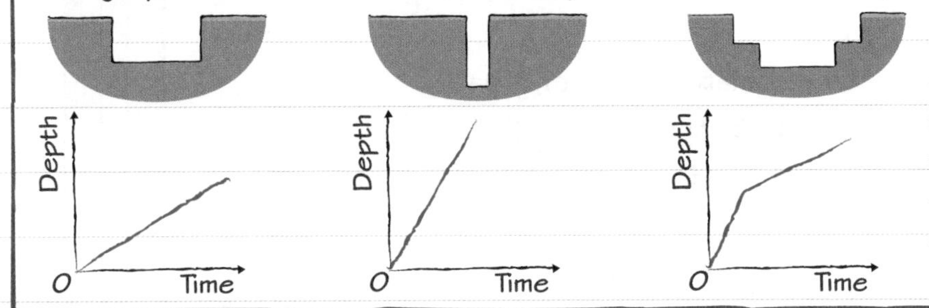

The narrower the pond, the faster the water depth will increase. In the last example, the water depth increases more slowly when the pond gets wider.

Now try this

target F

Here is a conversion graph.

(a) Use the graph to convert 6 kilograms to pounds. Give your answer to the nearest pound. *(1 mark)*

(b) Use the graph to convert 18 pounds to kilograms. *(1 mark)*

(c) Use the graph to convert 40 kilograms to pounds. Show clearly how you obtained your answer. *(2 marks)*

Draw lines on your graph to show the values you are reading off.

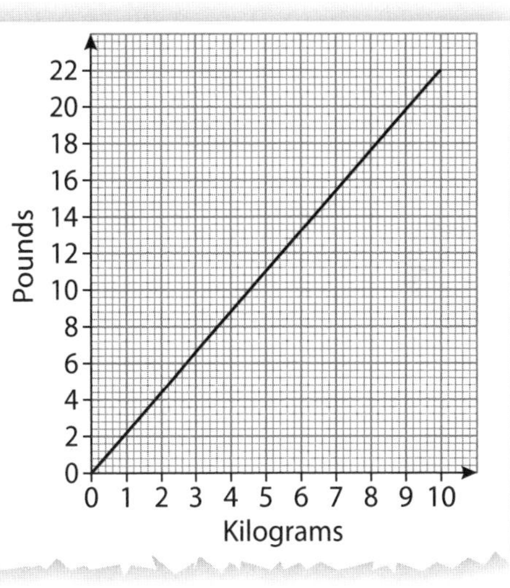

C D E F G

Measuring lines

You need to be able to use a ruler to draw and measure straight lines accurately.

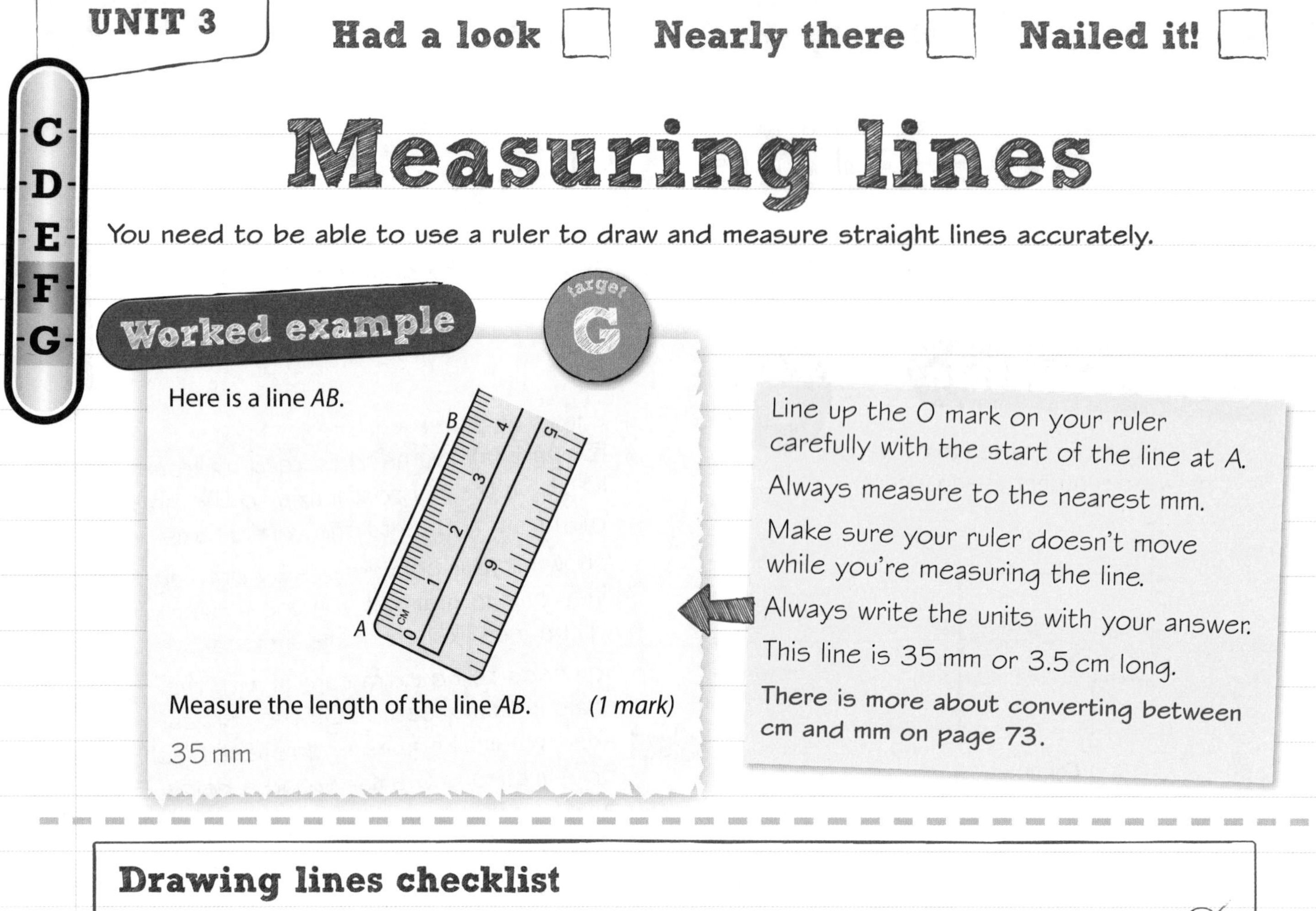

Worked example

target G

Here is a line AB.

Measure the length of the line AB. *(1 mark)*

35 mm

Line up the 0 mark on your ruler carefully with the start of the line at A.

Always measure to the nearest mm.

Make sure your ruler doesn't move while you're measuring the line.

Always write the units with your answer.

This line is 35 mm or 3.5 cm long.

There is more about converting between cm and mm on page 73.

Drawing lines checklist

- Check whether you are working in cm or mm. ✓
- Start the line at the 0 mark on your ruler. ✓
- Hold your ruler firmly. ✓
- Use a sharp pencil. ✓
- Draw to the nearest mm. ✓
- Label the length you have drawn. ✓

Estimating

You can use lengths that you know to estimate other lengths.

This diagram shows a man standing at the bottom of a cliff.

The man is 3 cm tall in the drawing and the cliff is 12 cm tall.
This means the cliff is 4 times as tall as the man.

A good estimate for the height of an adult male is 1.8 m.

$4 \times 1.8 = 7.2$

This means a good estimate for the height of the cliff is 7.2 m.

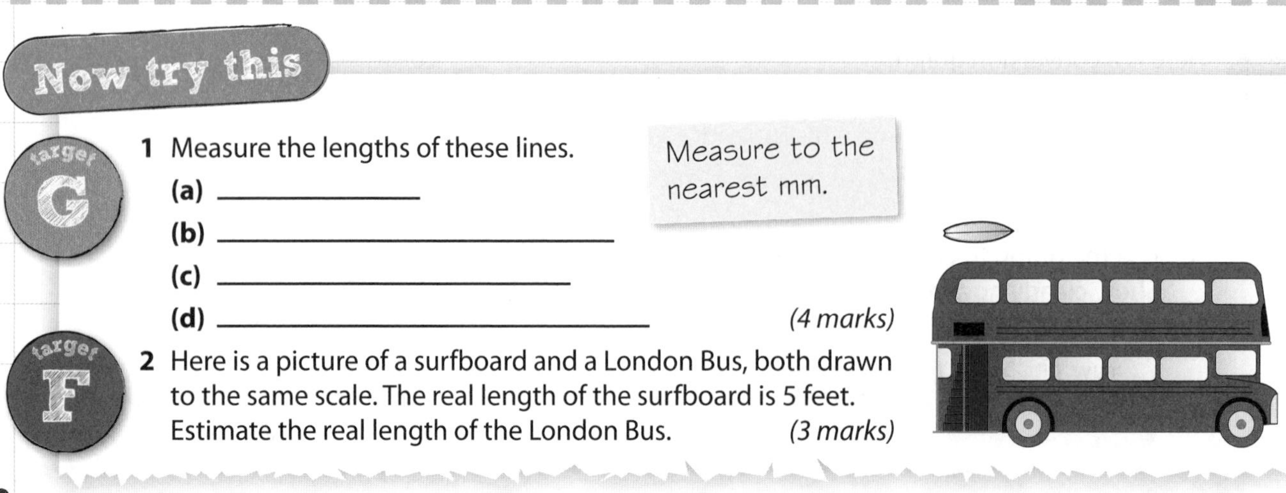

Now try this

target G

1 Measure the lengths of these lines.

(a) ______________

(b) ____________________________

(c) _________________________

(d) ____________________________ *(4 marks)*

Measure to the nearest mm.

target F

2 Here is a picture of a surfboard and a London Bus, both drawn to the same scale. The real length of the surfboard is 5 feet. Estimate the real length of the London Bus. *(3 marks)*

Metric units

Most of the units of measurement used in the UK are METRIC units.

You can convert between metric units by multiplying or dividing by 10, 100 or 1000.

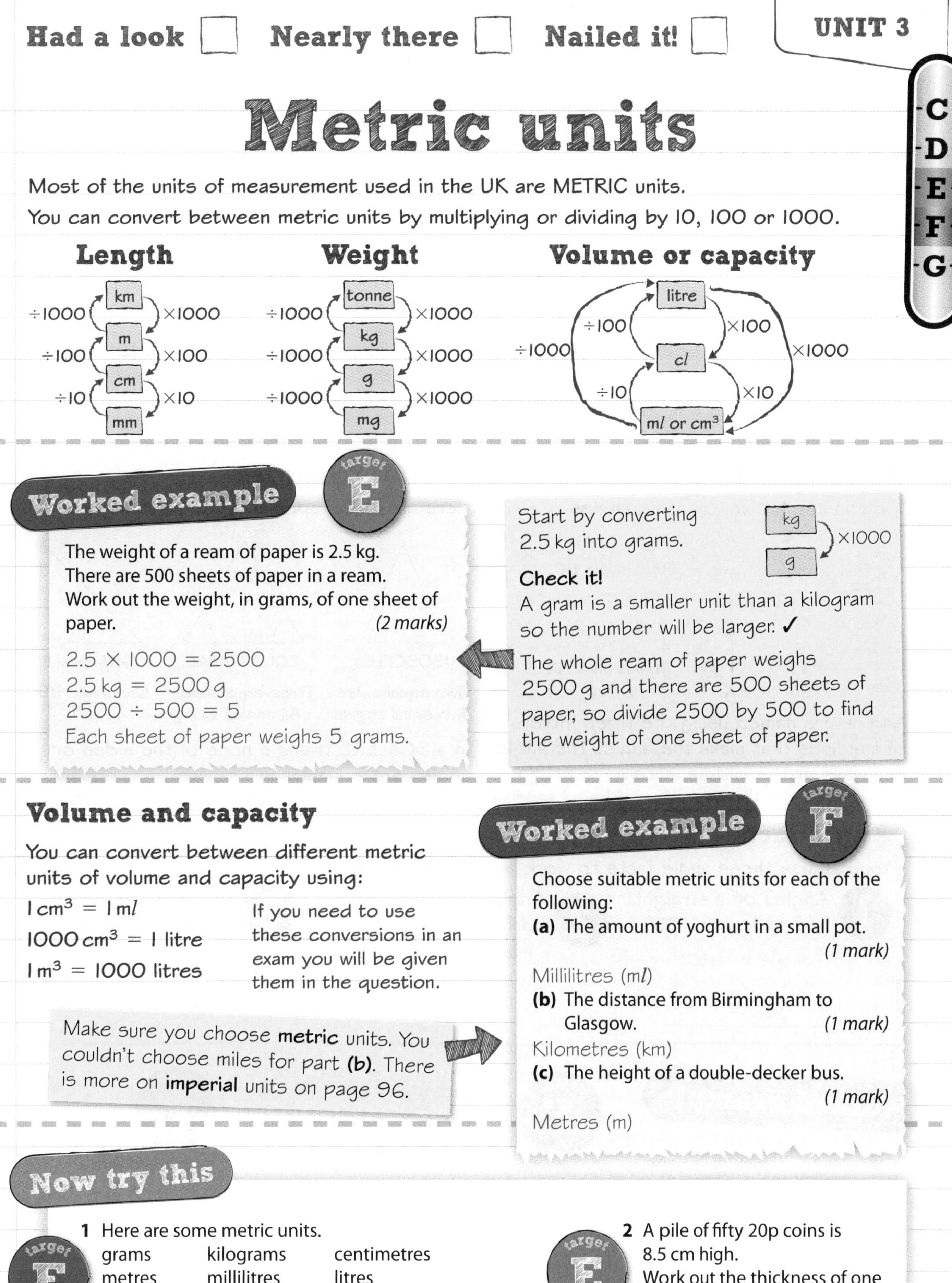

Worked example (target E)

The weight of a ream of paper is 2.5 kg.
There are 500 sheets of paper in a ream.
Work out the weight, in grams, of one sheet of paper. *(2 marks)*

$2.5 \times 1000 = 2500$

2.5 kg = 2500 g

$2500 \div 500 = 5$

Each sheet of paper weighs 5 grams.

Start by converting 2.5 kg into grams. (kg → g: ×1000)

Check it!

A gram is a smaller unit than a kilogram so the number will be larger. ✓

The whole ream of paper weighs 2500 g and there are 500 sheets of paper, so divide 2500 by 500 to find the weight of one sheet of paper.

Volume and capacity

You can convert between different metric units of volume and capacity using:

1 cm³ = 1 m*l*

1000 cm³ = 1 litre

1 m³ = 1000 litres

If you need to use these conversions in an exam you will be given them in the question.

Make sure you choose **metric** units. You couldn't choose miles for part **(b)**. There is more on **imperial** units on page 96.

Worked example (target F)

Choose suitable metric units for each of the following:

(a) The amount of yoghurt in a small pot. *(1 mark)*

Millilitres (m*l*)

(b) The distance from Birmingham to Glasgow. *(1 mark)*

Kilometres (km)

(c) The height of a double-decker bus. *(1 mark)*

Metres (m)

Now try this

(target F)

1 Here are some metric units.

grams kilograms centimetres
metres millilitres litres

From the above list, select the most suitable units to measure the following:

(a) the amount of lemonade in a large bottle *(1 mark)*
(b) the length of a lorry *(1 mark)*
(c) the weight of a large bag of sand *(1 mark)*
(d) the width of an envelope. *(1 mark)*

(target E)

2 A pile of fifty 20p coins is 8.5 cm high.
Work out the thickness of one 20p coin.
Give your answer in millimetres. *(2 marks)*

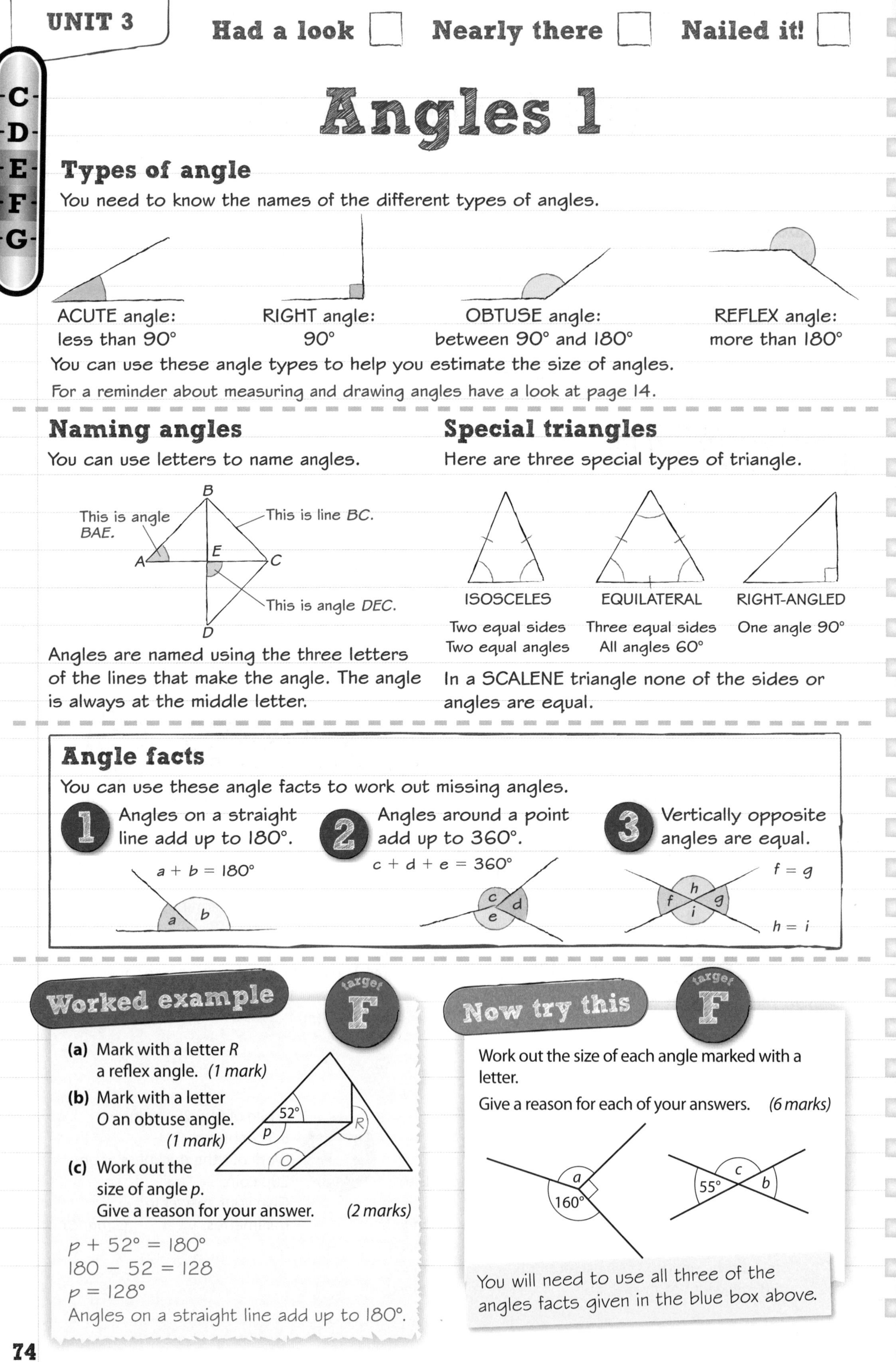

C D E F G

Angles 1

Types of angle

You need to know the names of the different types of angles.

ACUTE angle: less than 90°

RIGHT angle: 90°

OBTUSE angle: between 90° and 180°

REFLEX angle: more than 180°

You can use these angle types to help you estimate the size of angles.

For a reminder about measuring and drawing angles have a look at page 14.

Naming angles

You can use letters to name angles.

Angles are named using the three letters of the lines that make the angle. The angle is always at the middle letter.

Special triangles

Here are three special types of triangle.

ISOSCELES: Two equal sides, Two equal angles

EQUILATERAL: Three equal sides, All angles 60°

RIGHT-ANGLED: One angle 90°

In a SCALENE triangle none of the sides or angles are equal.

Angle facts

You can use these angle facts to work out missing angles.

1. Angles on a straight line add up to 180°. $a + b = 180°$
2. Angles around a point add up to 360°. $c + d + e = 360°$
3. Vertically opposite angles are equal. $f = g$, $h = i$

Worked example

target F

(a) Mark with a letter *R* a reflex angle. *(1 mark)*

(b) Mark with a letter *O* an obtuse angle. *(1 mark)*

(c) Work out the size of angle *p*. Give a reason for your answer. *(2 marks)*

$p + 52° = 180°$

$180 - 52 = 128$

$p = 128°$

Angles on a straight line add up to 180°.

Now try this

target F

Work out the size of each angle marked with a letter.

Give a reason for each of your answers. *(6 marks)*

You will need to use all three of the angles facts given in the blue box above.

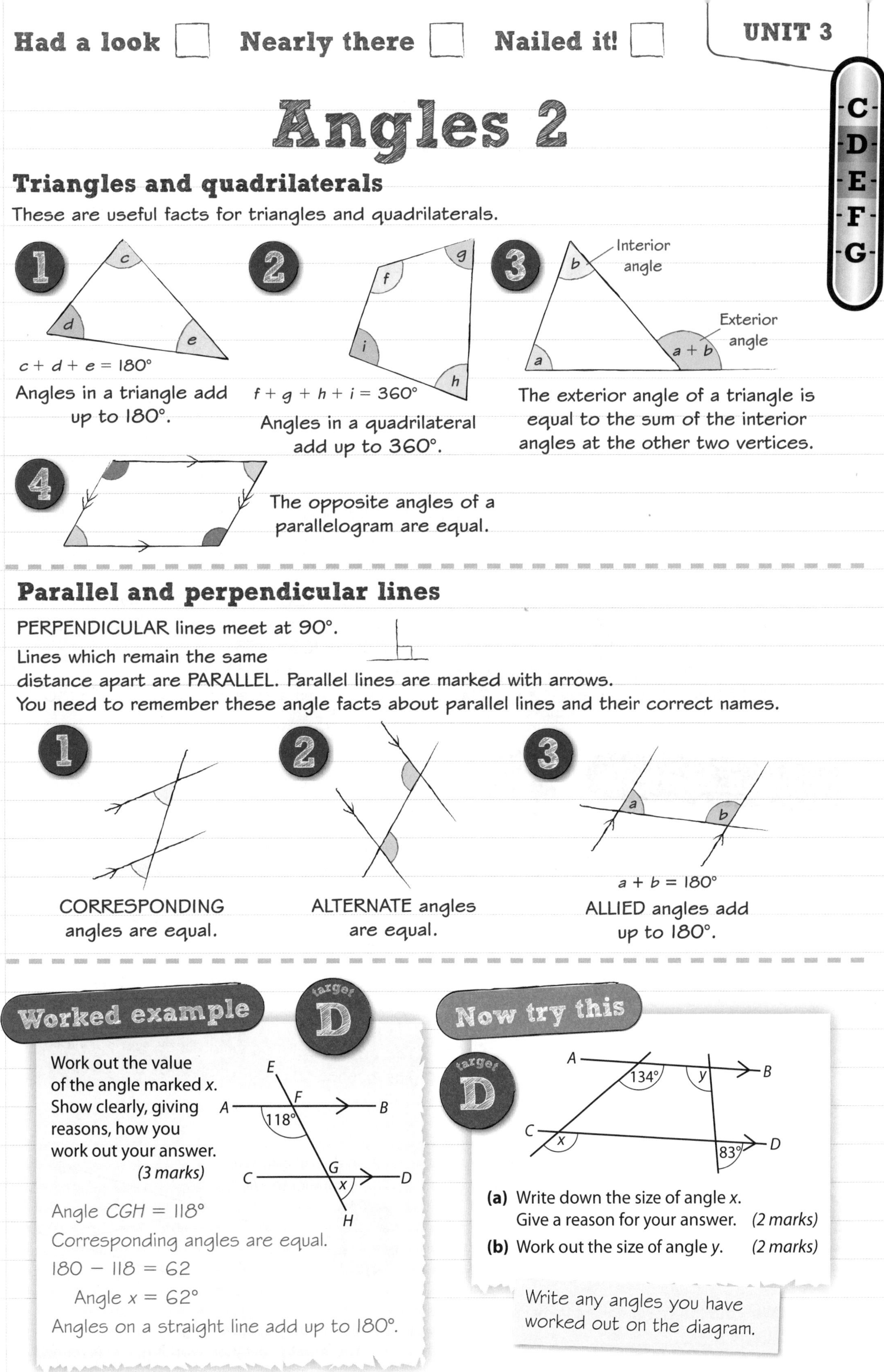

Angles 2

Triangles and quadrilaterals

These are useful facts for triangles and quadrilaterals.

1 $c + d + e = 180°$

Angles in a triangle add up to 180°.

2 $f + g + h + i = 360°$

Angles in a quadrilateral add up to 360°.

3 Interior angle b; Exterior angle $a + b$; angle a

The exterior angle of a triangle is equal to the sum of the interior angles at the other two vertices.

4 The opposite angles of a parallelogram are equal.

Parallel and perpendicular lines

PERPENDICULAR lines meet at 90°.

Lines which remain the same distance apart are PARALLEL. Parallel lines are marked with arrows.

You need to remember these angle facts about parallel lines and their correct names.

1 CORRESPONDING angles are equal.

2 ALTERNATE angles are equal.

3 $a + b = 180°$

ALLIED angles add up to 180°.

Worked example — target D

Work out the value of the angle marked x. Show clearly, giving reasons, how you work out your answer. *(3 marks)*

Angle $CGH = 118°$

Corresponding angles are equal.

$180 - 118 = 62$

Angle $x = 62°$

Angles on a straight line add up to 180°.

Now try this — target D

(a) Write down the size of angle x. Give a reason for your answer. *(2 marks)*

(b) Work out the size of angle y. *(2 marks)*

Write any angles you have worked out on the diagram.

Had a look ☐ Nearly there ☐ Nailed it! ☐

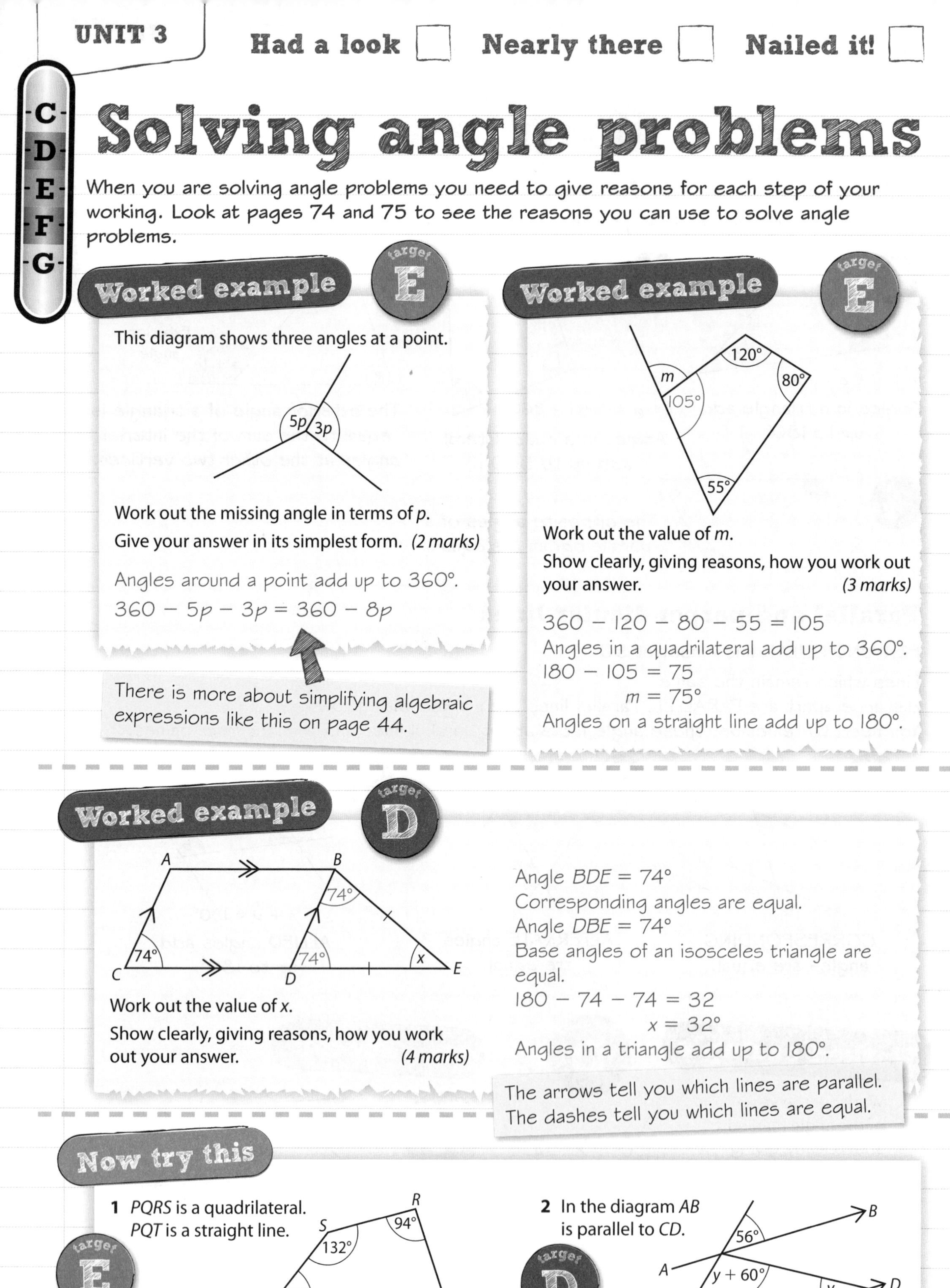

C D E F G

Solving angle problems

When you are solving angle problems you need to give reasons for each step of your working. Look at pages 74 and 75 to see the reasons you can use to solve angle problems.

Worked example (target E)

This diagram shows three angles at a point.

Work out the missing angle in terms of p.
Give your answer in its simplest form. *(2 marks)*

Angles around a point add up to 360°.
$360 - 5p - 3p = 360 - 8p$

There is more about simplifying algebraic expressions like this on page 44.

Worked example (target E)

Work out the value of m.
Show clearly, giving reasons, how you work out your answer. *(3 marks)*

$360 - 120 - 80 - 55 = 105$
Angles in a quadrilateral add up to 360°.
$180 - 105 = 75$
$m = 75°$
Angles on a straight line add up to 180°.

Worked example (target D)

Work out the value of x.
Show clearly, giving reasons, how you work out your answer. *(4 marks)*

Angle $BDE = 74°$
Corresponding angles are equal.
Angle $DBE = 74°$
Base angles of an isosceles triangle are equal.
$180 - 74 - 74 = 32$
$x = 32°$
Angles in a triangle add up to 180°.

The arrows tell you which lines are parallel.
The dashes tell you which lines are equal.

Now try this

1 (target E) $PQRS$ is a quadrilateral.
PQT is a straight line.

Work out the size of the angle marked x on the diagram. *(3 marks)*

2 (target D) In the diagram AB is parallel to CD.

Work out the size of angle y.
You **must** show your working. *(4 marks)*

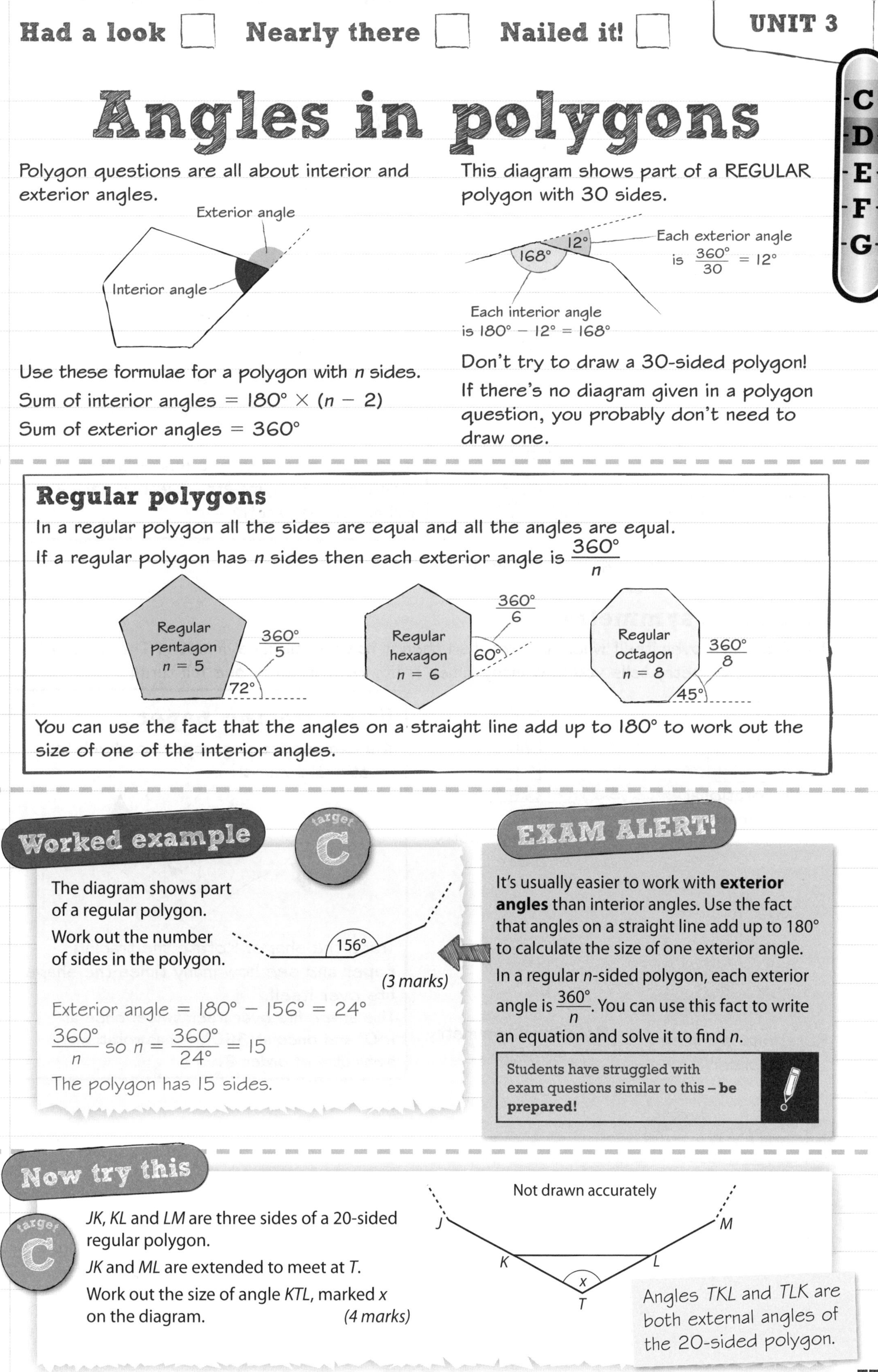

Angles in polygons

Polygon questions are all about interior and exterior angles.

This diagram shows part of a REGULAR polygon with 30 sides.

Use these formulae for a polygon with n sides.

Sum of interior angles $= 180° \times (n - 2)$

Sum of exterior angles $= 360°$

Don't try to draw a 30-sided polygon!

If there's no diagram given in a polygon question, you probably don't need to draw one.

Regular polygons

In a regular polygon all the sides are equal and all the angles are equal.

If a regular polygon has n sides then each exterior angle is $\frac{360°}{n}$

You can use the fact that the angles on a straight line add up to 180° to work out the size of one of the interior angles.

Worked example

target C

The diagram shows part of a regular polygon.

Work out the number of sides in the polygon.

(3 marks)

Exterior angle $= 180° - 156° = 24°$

$\frac{360°}{n}$ so $n = \frac{360°}{24°} = 15$

The polygon has 15 sides.

EXAM ALERT!

It's usually easier to work with **exterior angles** than interior angles. Use the fact that angles on a straight line add up to 180° to calculate the size of one exterior angle.

In a regular n-sided polygon, each exterior angle is $\frac{360°}{n}$. You can use this fact to write an equation and solve it to find n.

Students have struggled with exam questions similar to this – **be prepared!**

Now try this

target C

JK, KL and LM are three sides of a 20-sided regular polygon.

JK and ML are extended to meet at T.

Work out the size of angle KTL, marked x on the diagram. *(4 marks)*

Angles TKL and TLK are both external angles of the 20-sided polygon.

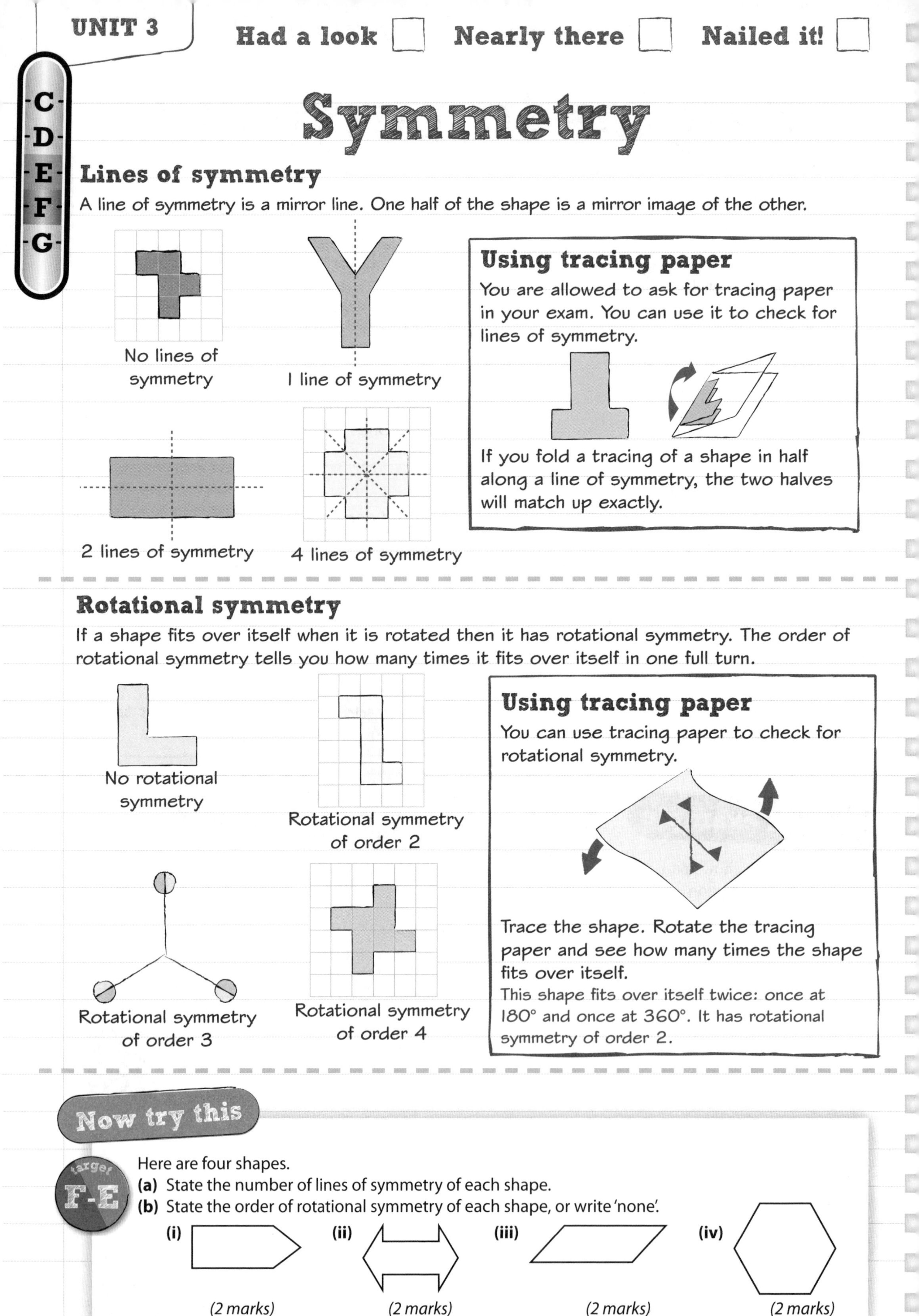

C D E F G

Symmetry

Lines of symmetry

A line of symmetry is a mirror line. One half of the shape is a mirror image of the other.

No lines of symmetry

1 line of symmetry

2 lines of symmetry

4 lines of symmetry

Using tracing paper

You are allowed to ask for tracing paper in your exam. You can use it to check for lines of symmetry.

If you fold a tracing of a shape in half along a line of symmetry, the two halves will match up exactly.

Rotational symmetry

If a shape fits over itself when it is rotated then it has rotational symmetry. The order of rotational symmetry tells you how many times it fits over itself in one full turn.

No rotational symmetry

Rotational symmetry of order 2

Rotational symmetry of order 3

Rotational symmetry of order 4

Using tracing paper

You can use tracing paper to check for rotational symmetry.

Trace the shape. Rotate the tracing paper and see how many times the shape fits over itself.

This shape fits over itself twice: once at 180° and once at 360°. It has rotational symmetry of order 2.

Now try this

target F-E

Here are four shapes.

(a) State the number of lines of symmetry of each shape.

(b) State the order of rotational symmetry of each shape, or write 'none'.

(i) *(2 marks)*

(ii) *(2 marks)*

(iii) *(2 marks)*

(iv) *(2 marks)*

C D E F G

Quadrilaterals

Quadrilaterals

A QUADRILATERAL is any four-sided shape. The lines joining the opposite corners of a quadrilateral are called DIAGONALS.

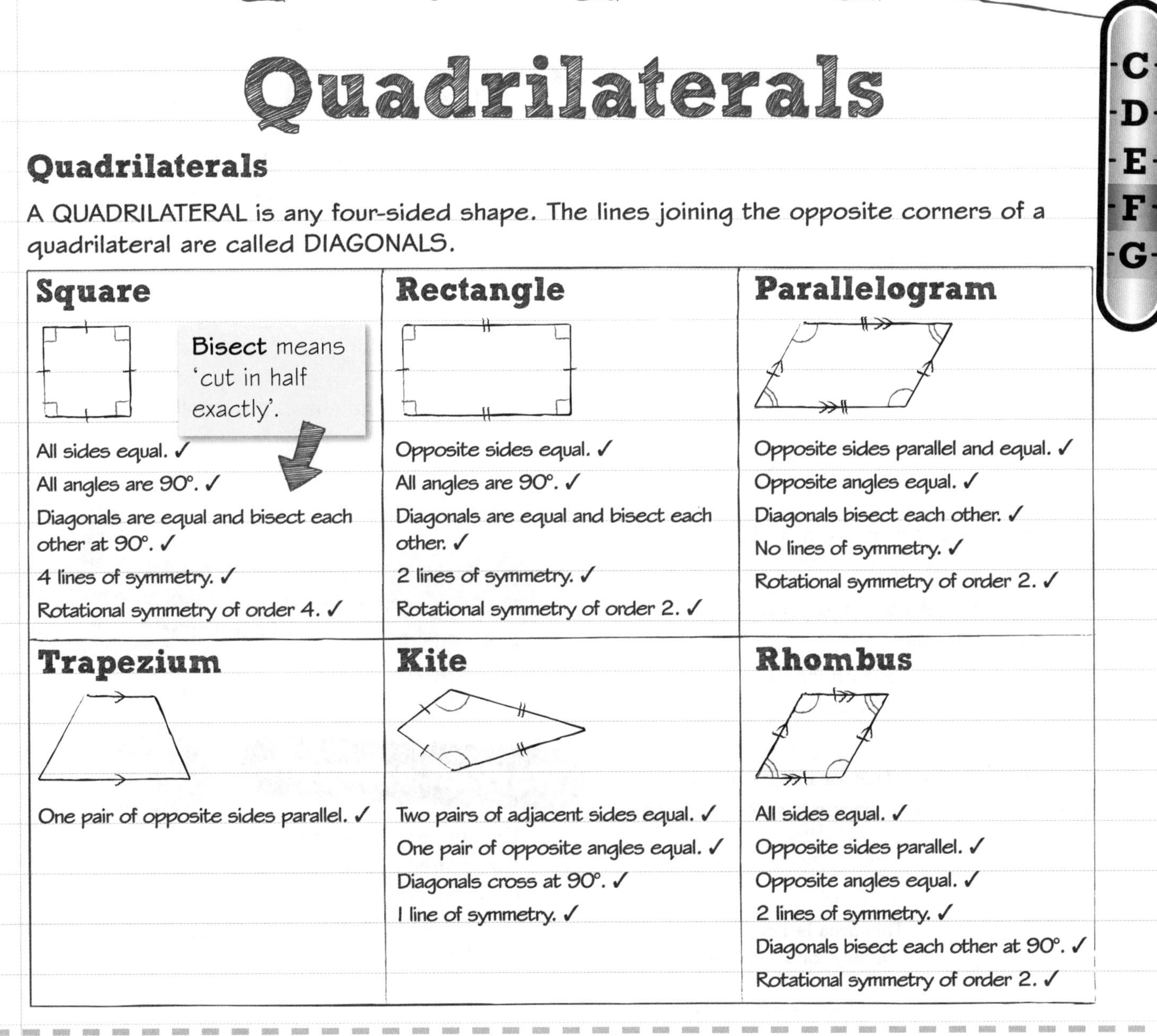

Now try this

Target G-F

Join the name of each 2-D shape to its correct description with a straight line. The first one has been done for you. *(5 marks)*

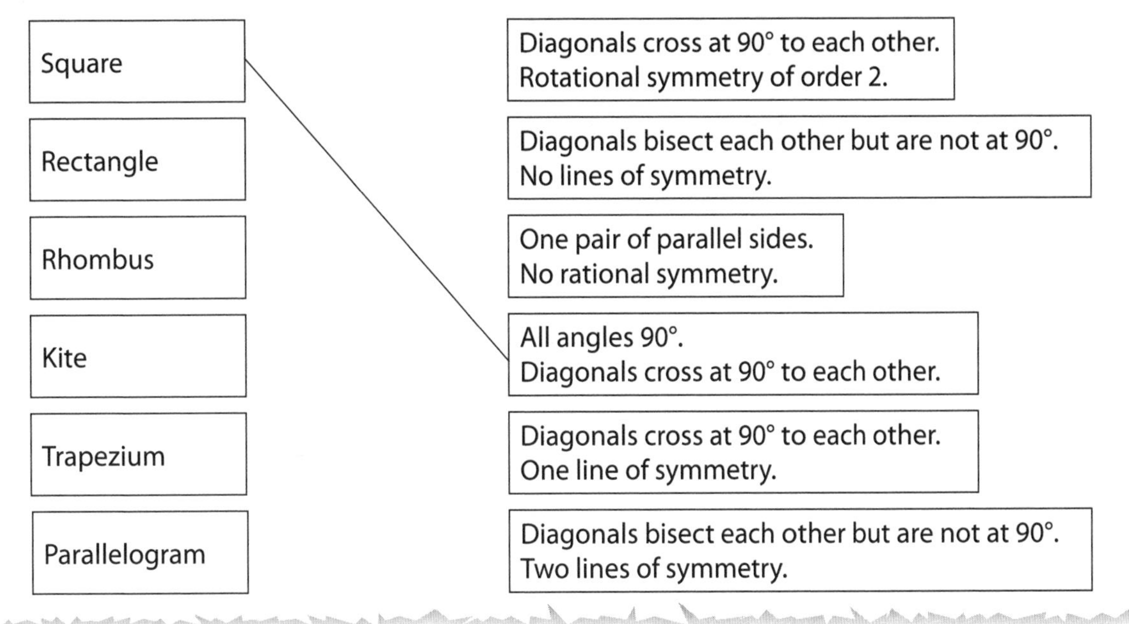

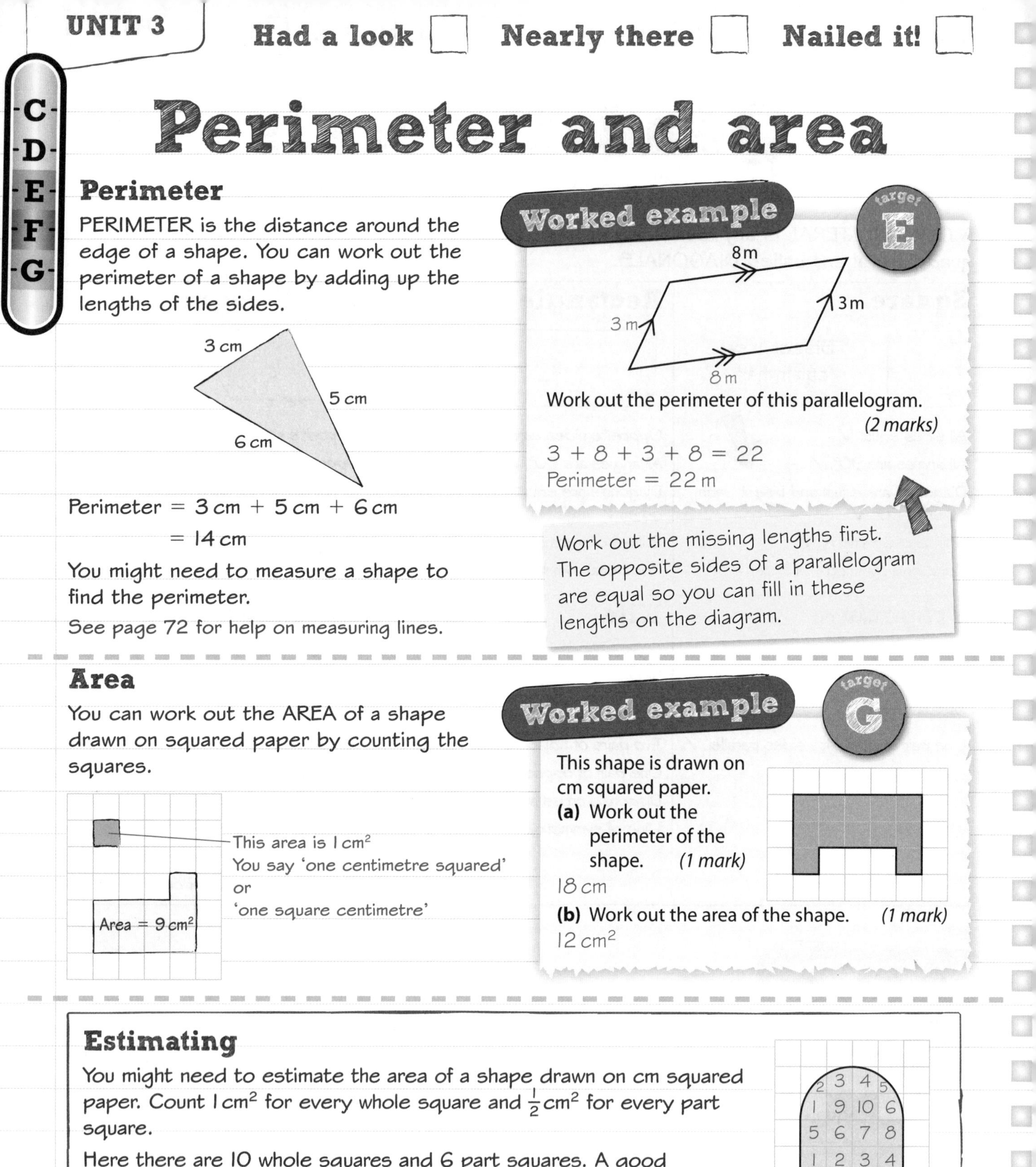

C D E F G

Perimeter and area

Perimeter

PERIMETER is the distance around the edge of a shape. You can work out the perimeter of a shape by adding up the lengths of the sides.

Perimeter = 3 cm + 5 cm + 6 cm
= 14 cm

You might need to measure a shape to find the perimeter.

See page 72 for help on measuring lines.

Worked example (target E)

Work out the perimeter of this parallelogram. *(2 marks)*

3 + 8 + 3 + 8 = 22
Perimeter = 22 m

Work out the missing lengths first. The opposite sides of a parallelogram are equal so you can fill in these lengths on the diagram.

Area

You can work out the AREA of a shape drawn on squared paper by counting the squares.

This area is 1 cm^2
You say 'one centimetre squared'
or
'one square centimetre'

Worked example (target G)

This shape is drawn on cm squared paper.

(a) Work out the perimeter of the shape. *(1 mark)*

18 cm

(b) Work out the area of the shape. *(1 mark)*

12 cm^2

Estimating

You might need to estimate the area of a shape drawn on cm squared paper. Count 1 cm^2 for every whole square and $\frac{1}{2}$ cm^2 for every part square.

Here there are 10 whole squares and 6 part squares. A good estimate is 13 cm^2.

Now try this

Estimate the area of this oval shape. Each square is 1 unit of area. *(2 marks)*

Count 1 unit for each whole square and half a unit for each part square.

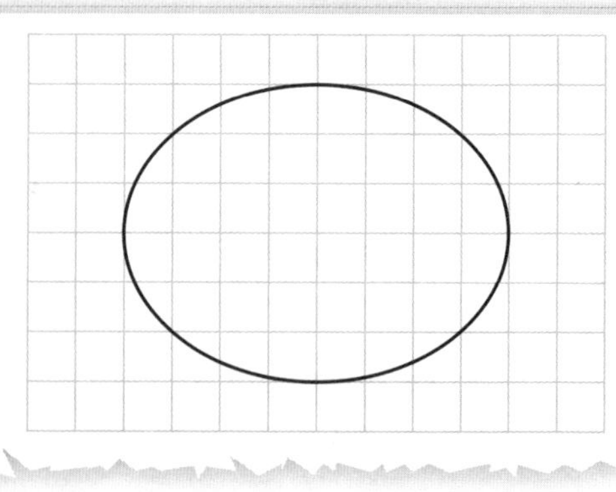

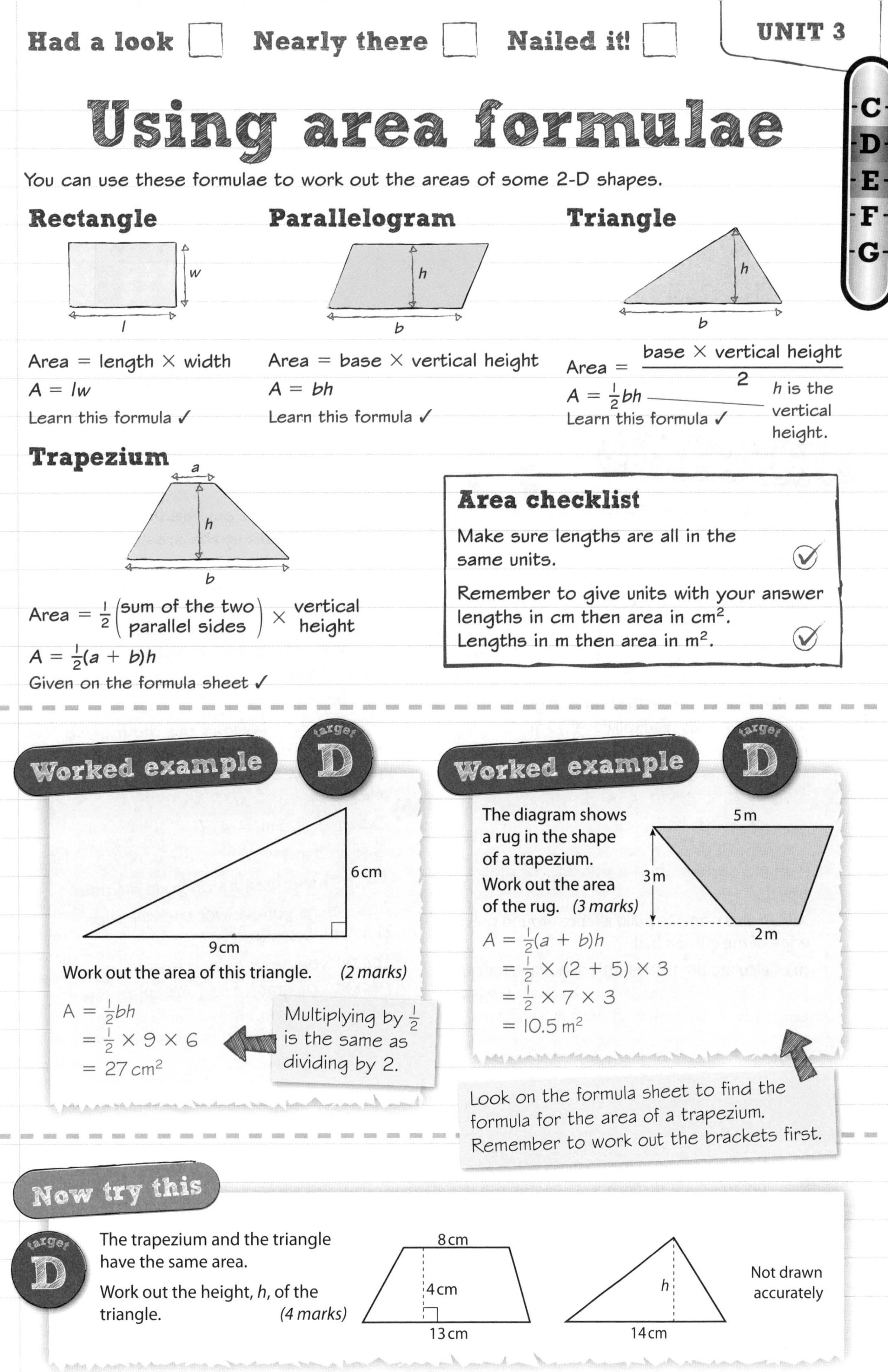

C D E F G

Using area formulae

You can use these formulae to work out the areas of some 2-D shapes.

Rectangle

Area = length × width

$A = lw$

Learn this formula ✓

Parallelogram

Area = base × vertical height

$A = bh$

Learn this formula ✓

Triangle

$$\text{Area} = \frac{\text{base} \times \text{vertical height}}{2}$$

$A = \frac{1}{2}bh$ — h is the vertical height.

Learn this formula ✓

Trapezium

$$\text{Area} = \frac{1}{2}\begin{pmatrix}\text{sum of the two} \\ \text{parallel sides}\end{pmatrix} \times \begin{matrix}\text{vertical} \\ \text{height}\end{matrix}$$

$A = \frac{1}{2}(a + b)h$

Given on the formula sheet ✓

Area checklist

Make sure lengths are all in the same units. ✓

Remember to give units with your answer lengths in cm then area in cm^2. Lengths in m then area in m^2. ✓

Worked example (target D)

Work out the area of this triangle. *(2 marks)*

$A = \frac{1}{2}bh$

$= \frac{1}{2} \times 9 \times 6$

$= 27\ \text{cm}^2$

Multiplying by $\frac{1}{2}$ is the same as dividing by 2.

Worked example (target D)

The diagram shows a rug in the shape of a trapezium. Work out the area of the rug. *(3 marks)*

$A = \frac{1}{2}(a + b)h$

$= \frac{1}{2} \times (2 + 5) \times 3$

$= \frac{1}{2} \times 7 \times 3$

$= 10.5\ \text{m}^2$

Look on the formula sheet to find the formula for the area of a trapezium. Remember to work out the brackets first.

Now try this (target D)

The trapezium and the triangle have the same area.

Work out the height, h, of the triangle. *(4 marks)*

C D E F G

Solving area problems

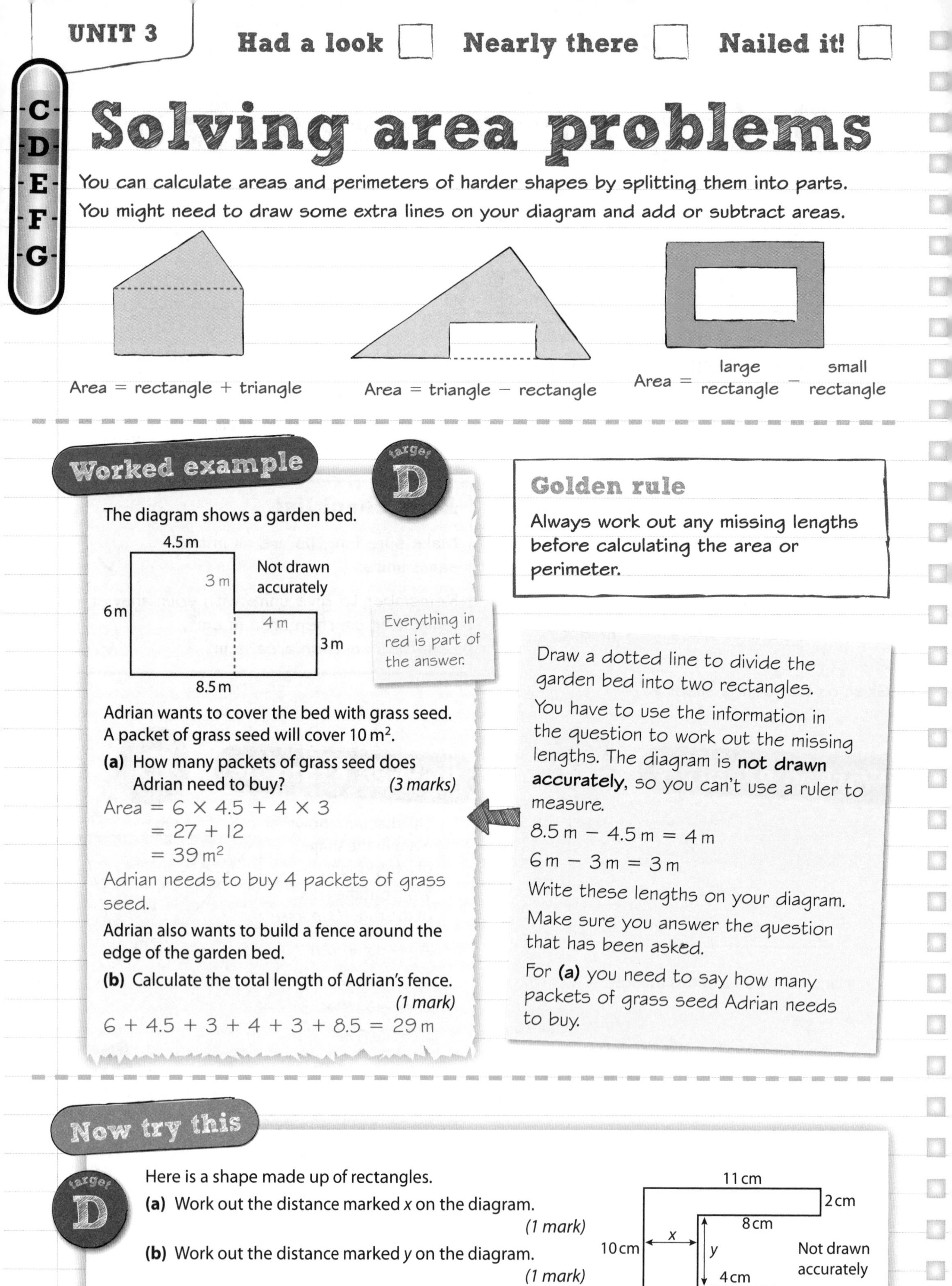

You can calculate areas and perimeters of harder shapes by splitting them into parts.
You might need to draw some extra lines on your diagram and add or subtract areas.

Area = rectangle + triangle

Area = triangle − rectangle

Area = large rectangle − small rectangle

Worked example

target D

The diagram shows a garden bed.

Everything in red is part of the answer.

Adrian wants to cover the bed with grass seed. A packet of grass seed will cover 10 m².

(a) How many packets of grass seed does Adrian need to buy? *(3 marks)*

Area $= 6 \times 4.5 + 4 \times 3$

$= 27 + 12$

$= 39\text{ m}^2$

Adrian needs to buy 4 packets of grass seed.

Adrian also wants to build a fence around the edge of the garden bed.

(b) Calculate the total length of Adrian's fence. *(1 mark)*

$6 + 4.5 + 3 + 4 + 3 + 8.5 = 29\text{ m}$

Golden rule

Always work out any missing lengths before calculating the area or perimeter.

Draw a dotted line to divide the garden bed into two rectangles.
You have to use the information in the question to work out the missing lengths. The diagram is **not drawn accurately**, so you can't use a ruler to measure.

$8.5\text{ m} - 4.5\text{ m} = 4\text{ m}$

$6\text{ m} - 3\text{ m} = 3\text{ m}$

Write these lengths on your diagram.
Make sure you answer the question that has been asked.
For **(a)** you need to say how many packets of grass seed Adrian needs to buy.

Now try this

target D

Here is a shape made up of rectangles.

(a) Work out the distance marked x on the diagram. *(1 mark)*

(b) Work out the distance marked y on the diagram. *(1 mark)*

(c) Work out the area of this shape. *(3 marks)*

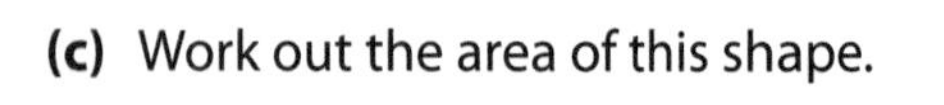

(d) Work out the perimeter of this shape. *(2 marks)*

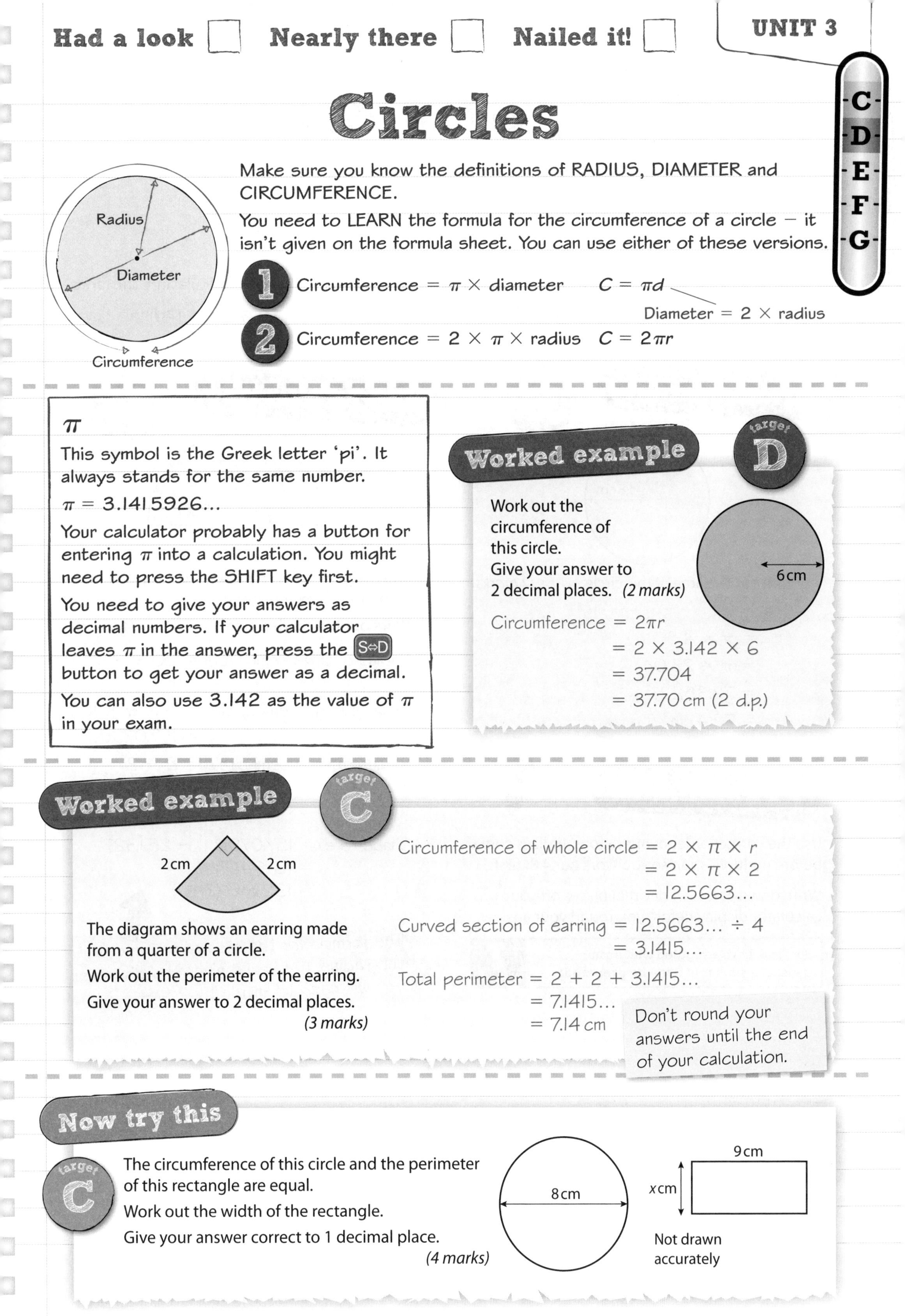

Circles

Radius

Diameter

Circumference

Make sure you know the definitions of RADIUS, DIAMETER and CIRCUMFERENCE.

You need to LEARN the formula for the circumference of a circle – it isn't given on the formula sheet. You can use either of these versions.

1. Circumference = $\pi \times$ diameter $C = \pi d$

   Diameter = 2 × radius

2. Circumference = $2 \times \pi \times$ radius $C = 2\pi r$

π

This symbol is the Greek letter 'pi'. It always stands for the same number.

$\pi = 3.1415926...$

Your calculator probably has a button for entering π into a calculation. You might need to press the SHIFT key first.

You need to give your answers as decimal numbers. If your calculator leaves π in the answer, press the S⇔D button to get your answer as a decimal.

You can also use 3.142 as the value of π in your exam.

Worked example

target D

Work out the circumference of this circle.

Give your answer to 2 decimal places. *(2 marks)*

6 cm

Circumference $= 2\pi r$

$= 2 \times 3.142 \times 6$

$= 37.704$

$= 37.70$ cm (2 d.p.)

Worked example

target C

2 cm 2 cm

The diagram shows an earring made from a quarter of a circle.

Work out the perimeter of the earring.

Give your answer to 2 decimal places.

(3 marks)

Circumference of whole circle $= 2 \times \pi \times r$

$= 2 \times \pi \times 2$

$= 12.5663...$

Curved section of earring $= 12.5663... \div 4$

$= 3.1415...$

Total perimeter $= 2 + 2 + 3.1415...$

$= 7.1415...$

$= 7.14$ cm

Don't round your answers until the end of your calculation.

Now try this

target C

The circumference of this circle and the perimeter of this rectangle are equal.

Work out the width of the rectangle.

Give your answer correct to 1 decimal place.

(4 marks)

8 cm

9 cm

x cm

Not drawn accurately

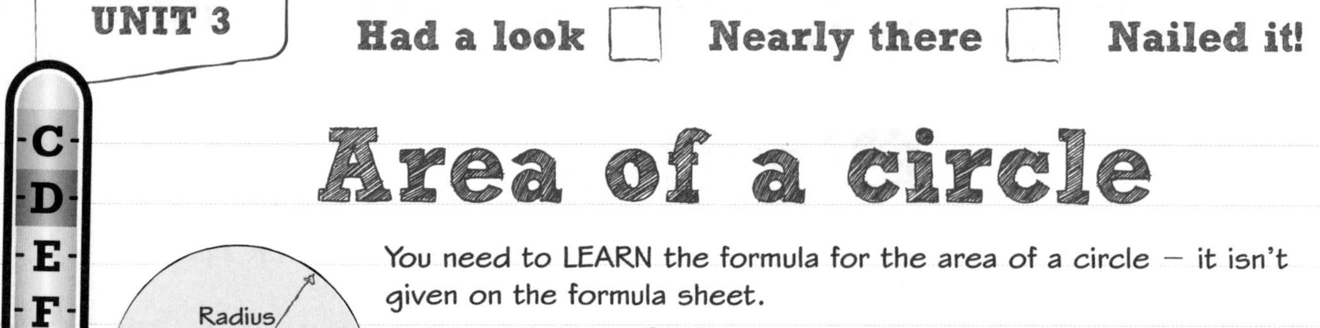

Area of a circle

You need to LEARN the formula for the area of a circle – it isn't given on the formula sheet.

Area = $\pi \times \text{radius}^2$

$A = \pi \times r \times r = \pi r^2$

You always need to use the RADIUS when you are calculating the area.

If you are given the diameter, divide it by 2 to get the radius.

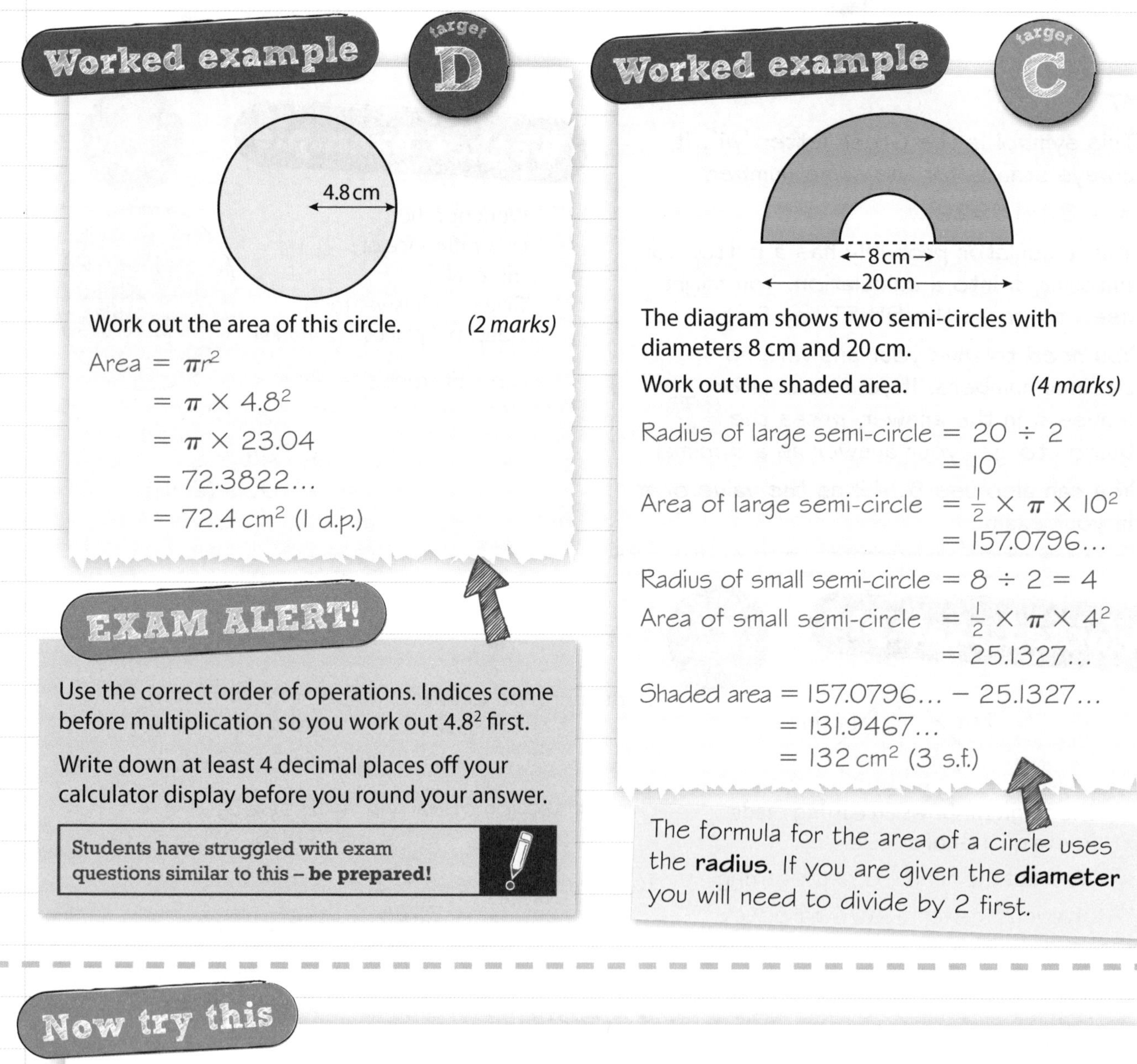

Worked example (target D)

Work out the area of this circle. *(2 marks)*

Area $= \pi r^2$
$= \pi \times 4.8^2$
$= \pi \times 23.04$
$= 72.3822...$
$= 72.4\ \text{cm}^2$ (1 d.p.)

EXAM ALERT!

Use the correct order of operations. Indices come before multiplication so you work out 4.8^2 first.

Write down at least 4 decimal places off your calculator display before you round your answer.

Students have struggled with exam questions similar to this – **be prepared!**

Worked example (target C)

The diagram shows two semi-circles with diameters 8 cm and 20 cm.

Work out the shaded area. *(4 marks)*

Radius of large semi-circle $= 20 \div 2 = 10$

Area of large semi-circle $= \frac{1}{2} \times \pi \times 10^2 = 157.0796...$

Radius of small semi-circle $= 8 \div 2 = 4$

Area of small semi-circle $= \frac{1}{2} \times \pi \times 4^2 = 25.1327...$

Shaded area $= 157.0796... - 25.1327...$
$= 131.9467...$
$= 132\ \text{cm}^2$ (3 s.f.)

The formula for the area of a circle uses the **radius**. If you are given the **diameter** you will need to divide by 2 first.

Now try this

This circle and this square are equal in area.

Work out the length of the side of the square.

Give your answer correct to 1 decimal place. *(4 marks)*

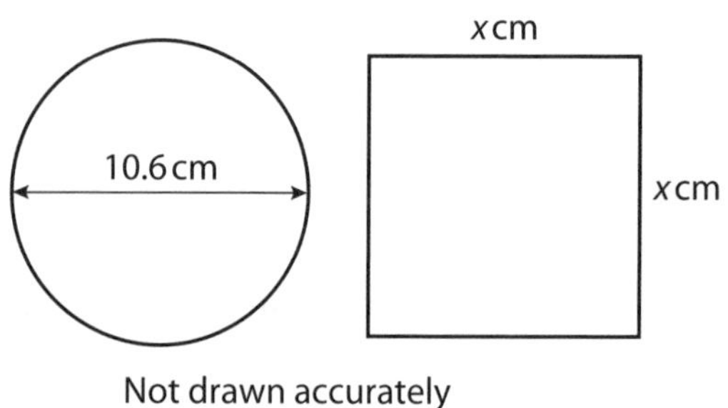

Work out the area of the circle. This is the same as the area of the square. Work out the square root to find x.

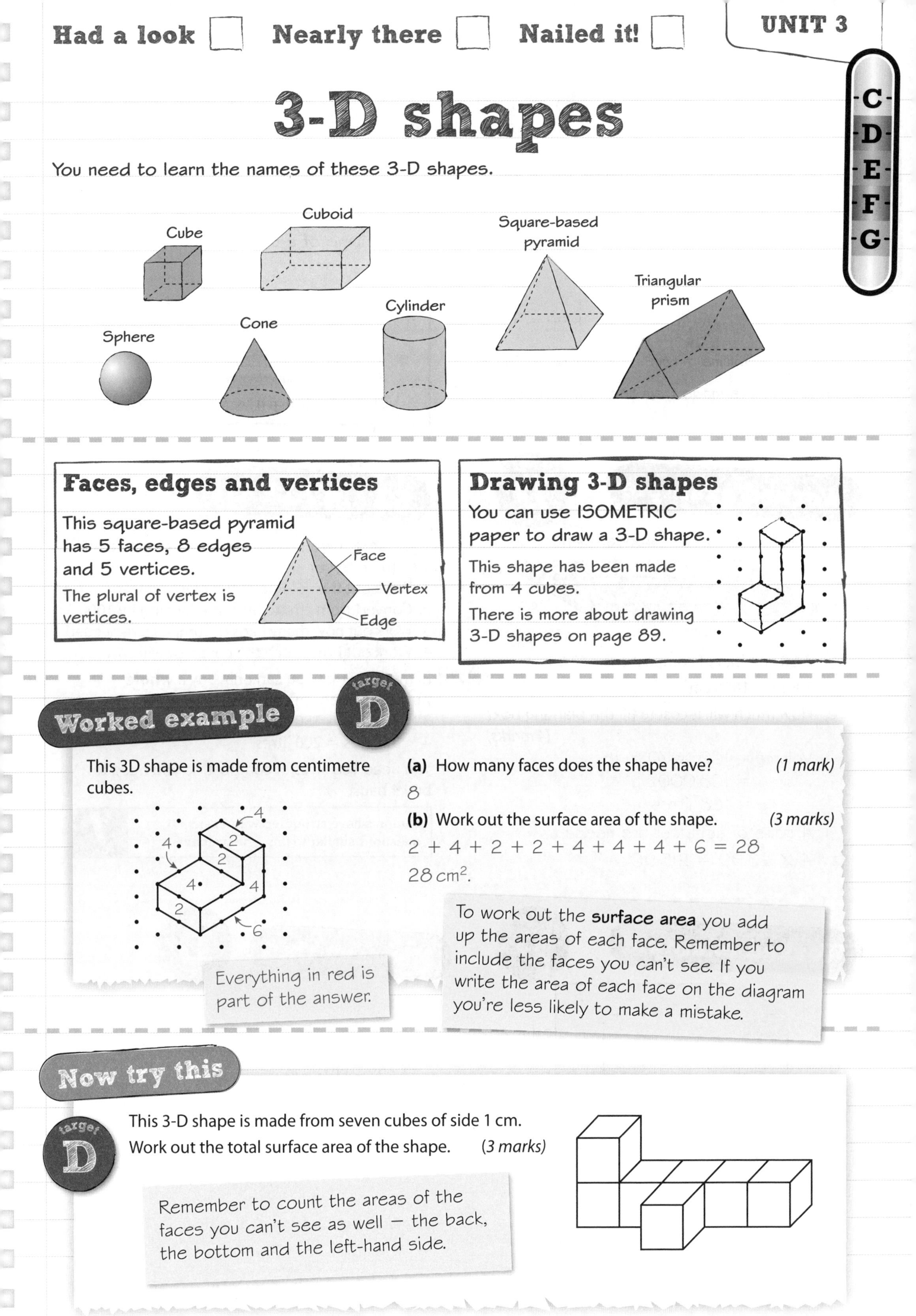

3-D shapes

You need to learn the names of these 3-D shapes.

Cube, Cuboid, Square-based pyramid, Triangular prism, Sphere, Cone, Cylinder

Faces, edges and vertices

This square-based pyramid has 5 faces, 8 edges and 5 vertices.

The plural of vertex is vertices.

Drawing 3-D shapes

You can use ISOMETRIC paper to draw a 3-D shape.

This shape has been made from 4 cubes.

There is more about drawing 3-D shapes on page 89.

Worked example (target D)

This 3D shape is made from centimetre cubes.

(a) How many faces does the shape have? *(1 mark)*

8

(b) Work out the surface area of the shape. *(3 marks)*

$2 + 4 + 2 + 2 + 4 + 4 + 4 + 6 = 28$

28 cm^2.

Everything in red is part of the answer.

To work out the **surface area** you add up the areas of each face. Remember to include the faces you can't see. If you write the area of each face on the diagram you're less likely to make a mistake.

Now try this (target D)

This 3-D shape is made from seven cubes of side 1 cm.

Work out the total surface area of the shape. *(3 marks)*

Remember to count the areas of the faces you can't see as well – the back, the bottom and the left-hand side.

C D E F G

Volume

The VOLUME of a 3-D shape is the amount of space it takes up.

The most common units of volume are cm^3 or m^3.

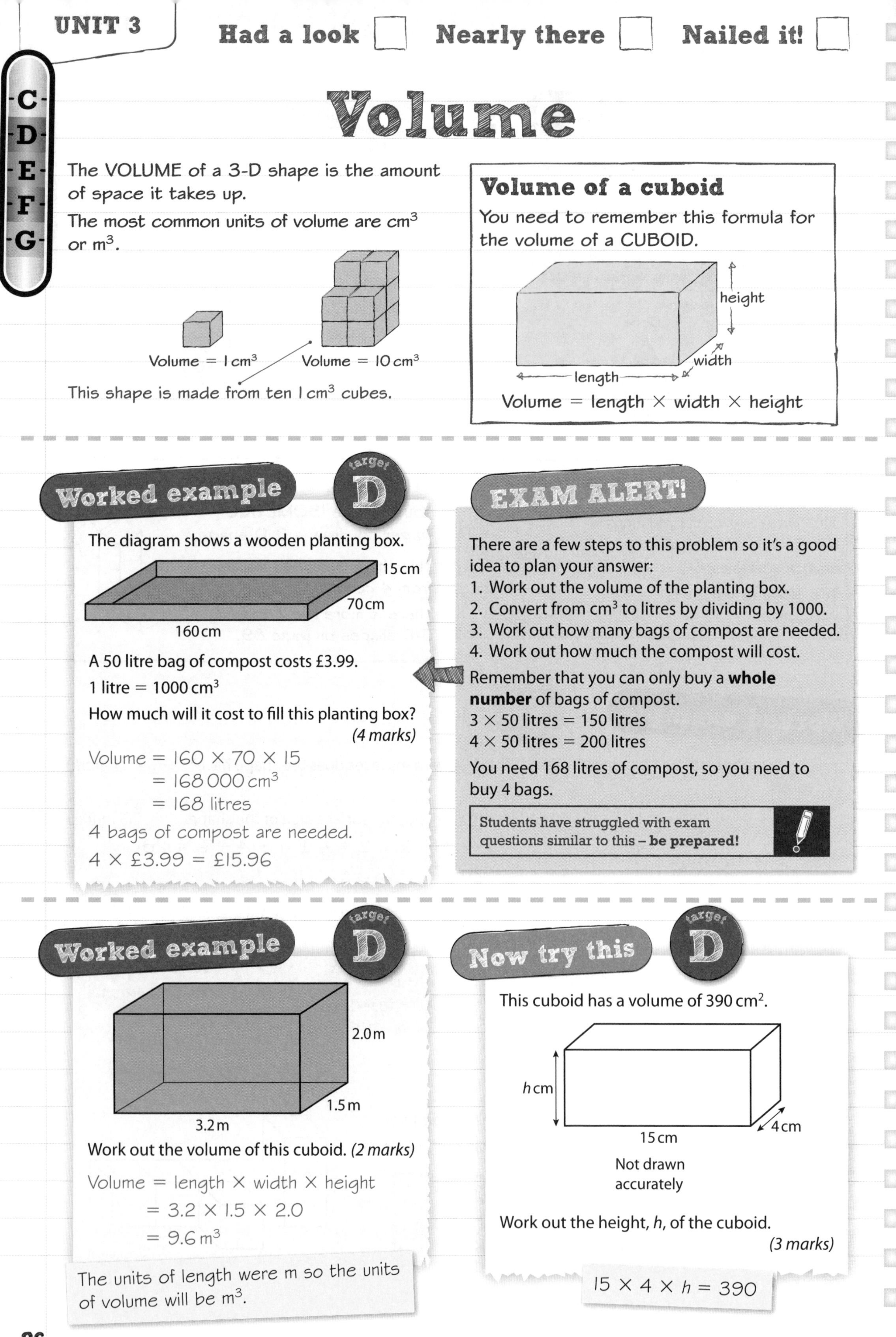

Volume of a cuboid

You need to remember this formula for the volume of a CUBOID.

Volume = length × width × height

Worked example (target D)

The diagram shows a wooden planting box.

15 cm, 70 cm, 160 cm

A 50 litre bag of compost costs £3.99.

1 litre = 1000 cm^3

How much will it cost to fill this planting box? *(4 marks)*

Volume = 160 × 70 × 15
= 168 000 cm^3
= 168 litres

4 bags of compost are needed.

4 × £3.99 = £15.96

EXAM ALERT!

There are a few steps to this problem so it's a good idea to plan your answer:

1. Work out the volume of the planting box.
2. Convert from cm^3 to litres by dividing by 1000.
3. Work out how many bags of compost are needed.
4. Work out how much the compost will cost.

Remember that you can only buy a **whole number** of bags of compost.

3 × 50 litres = 150 litres
4 × 50 litres = 200 litres

You need 168 litres of compost, so you need to buy 4 bags.

Students have struggled with exam questions similar to this – **be prepared!**

Worked example (target D)

2.0 m, 1.5 m, 3.2 m

Work out the volume of this cuboid. *(2 marks)*

Volume = length × width × height
= 3.2 × 1.5 × 2.0
= 9.6 m^3

The units of length were m so the units of volume will be m^3.

Now try this (target D)

This cuboid has a volume of 390 cm^2.

h cm, 15 cm, 4 cm

Not drawn accurately

Work out the height, h, of the cuboid. *(3 marks)*

$15 \times 4 \times h = 390$

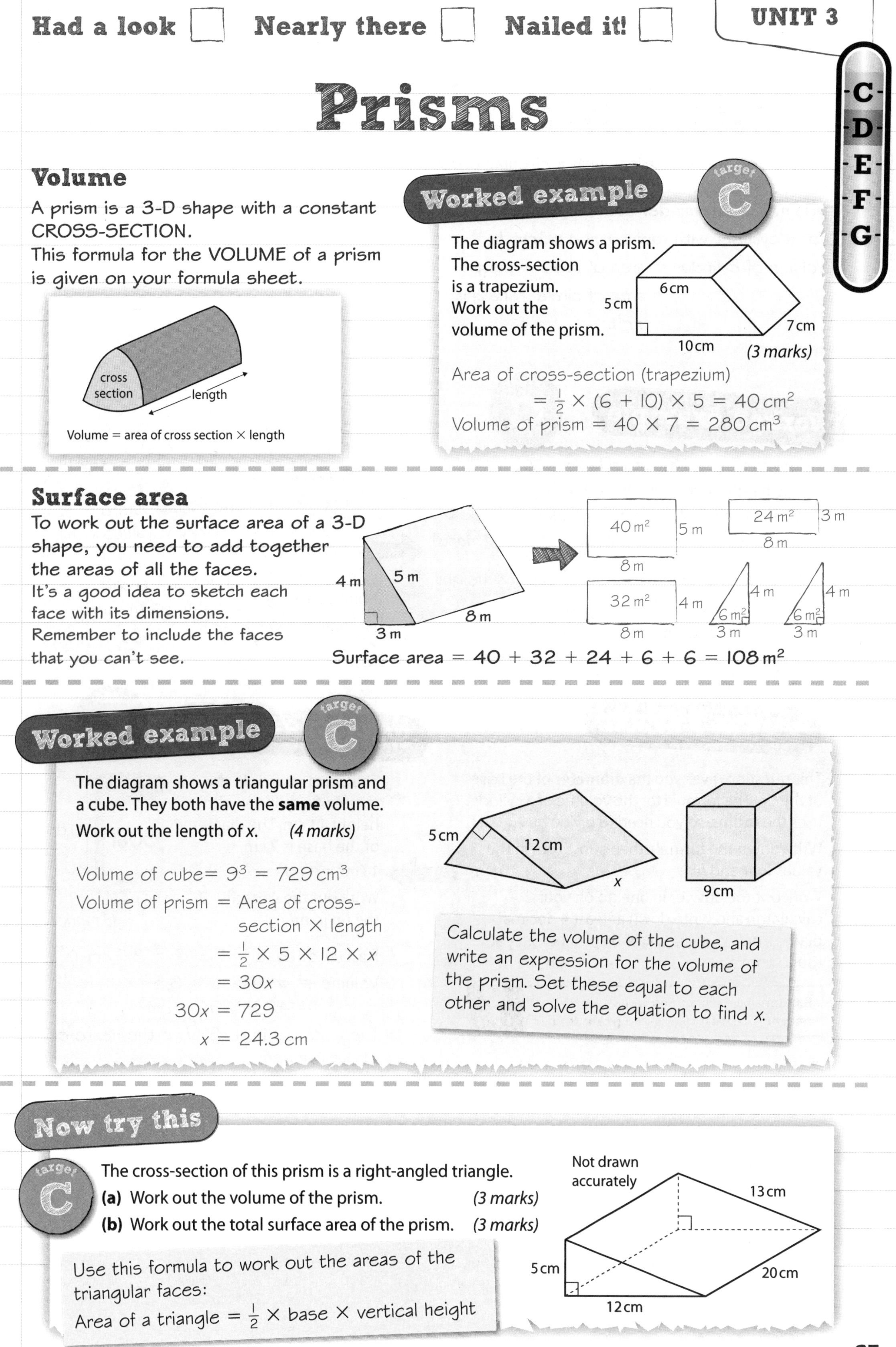

C D E F G

Prisms

Volume

A prism is a 3-D shape with a constant CROSS-SECTION.
This formula for the VOLUME of a prism is given on your formula sheet.

cross section
length

Volume = area of cross section × length

Worked example

target C

The diagram shows a prism. The cross-section is a trapezium. Work out the volume of the prism. *(3 marks)*

6 cm
5 cm
7 cm
10 cm

Area of cross-section (trapezium)

$= \frac{1}{2} \times (6 + 10) \times 5 = 40\,\text{cm}^2$

Volume of prism $= 40 \times 7 = 280\,\text{cm}^3$

Surface area

To work out the surface area of a 3-D shape, you need to add together the areas of all the faces.

It's a good idea to sketch each face with its dimensions.

Remember to include the faces that you can't see.

4 m
5 m
8 m
3 m

$40\,\text{m}^2$ (8 m × 5 m)
$24\,\text{m}^2$ (8 m × 3 m)
$32\,\text{m}^2$ (8 m × 4 m)
$6\,\text{m}^2$ (3 m, 4 m)
$6\,\text{m}^2$ (3 m, 4 m)

Surface area $= 40 + 32 + 24 + 6 + 6 = 108\,\text{m}^2$

Worked example

target C

The diagram shows a triangular prism and a cube. They both have the **same** volume.
Work out the length of x. *(4 marks)*

5 cm
12 cm
x
9 cm

Volume of cube $= 9^3 = 729\,\text{cm}^3$

Volume of prism = Area of cross-section × length

$= \frac{1}{2} \times 5 \times 12 \times x$

$= 30x$

$30x = 729$

$x = 24.3\,\text{cm}$

Calculate the volume of the cube, and write an expression for the volume of the prism. Set these equal to each other and solve the equation to find x.

Now try this

target C

The cross-section of this prism is a right-angled triangle.

(a) Work out the volume of the prism. *(3 marks)*

(b) Work out the total surface area of the prism. *(3 marks)*

Not drawn accurately

13 cm
5 cm
20 cm
12 cm

Use this formula to work out the areas of the triangular faces:
Area of a triangle $= \frac{1}{2} \times$ base $\times$ vertical height

C D E F G

Volume of a cylinder

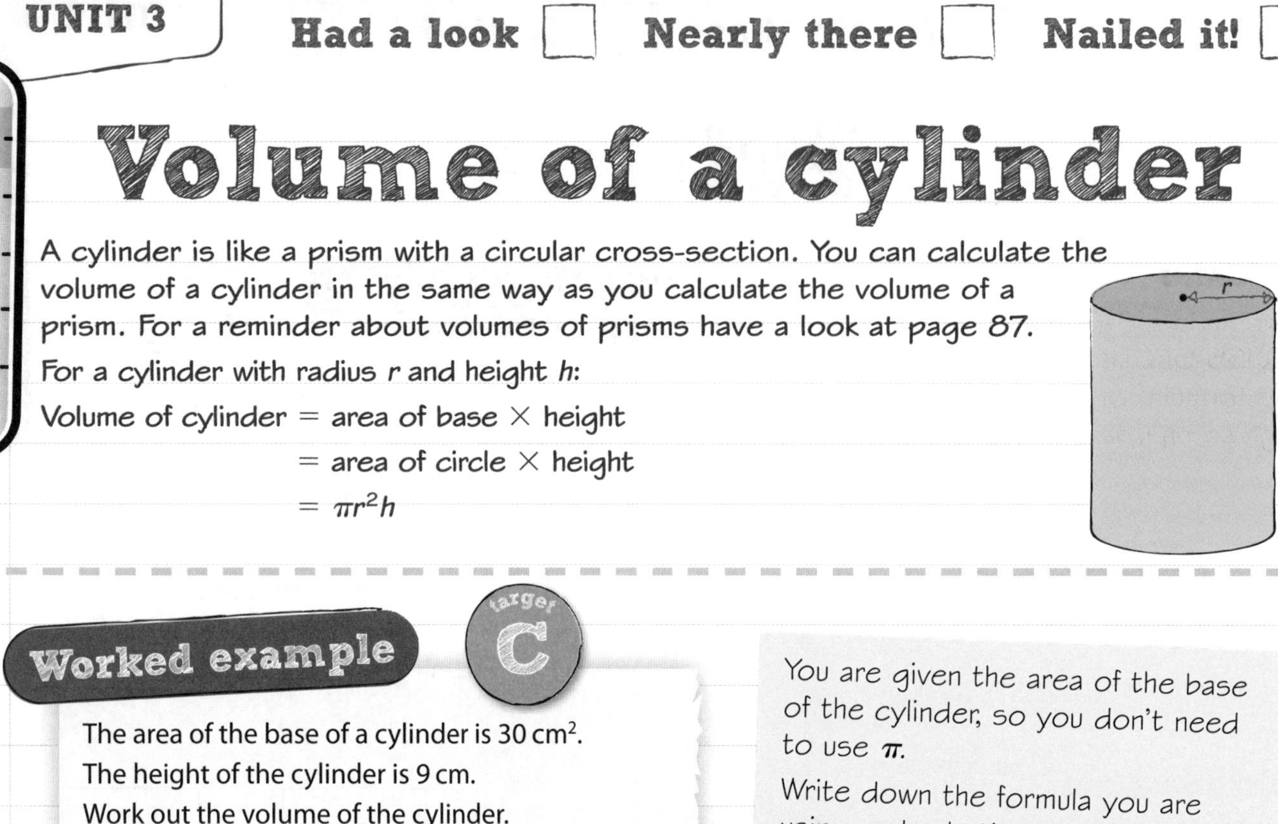

A cylinder is like a prism with a circular cross-section. You can calculate the volume of a cylinder in the same way as you calculate the volume of a prism. For a reminder about volumes of prisms have a look at page 87.

For a cylinder with radius r and height h:

Volume of cylinder = area of base × height
= area of circle × height
= $\pi r^2 h$

Worked example (target C)

The area of the base of a cylinder is 30 cm².
The height of the cylinder is 9 cm.
Work out the volume of the cylinder.
Give the units with your answer. *(2 marks)*

Volume of cylinder = area of base × height
= 30 × 9
= 270 cm³

You are given the area of the base of the cylinder, so you don't need to use π.

Write down the formula you are using and substitute the values you are given.

The units of area are cm² and the units of length are cm, so the units of volume will be cm³.

EXAM ALERT!

This question gives you the **diameter** of the base of the tin. The formula for the volume of a cylinder uses the **radius**, so you need to divide by 2.

Write down the formula then substitute in the values for r and h.

Work out the answer in one go on your calculator, and write down at least 4 decimal places from your calculator display before you round your answer.

Students have struggled with exam questions similar to this – **be prepared!**

Worked example (target C)

This tin of soup is in the shape of a cylinder with height 11 cm. The diameter of the base is 7 cm.

1 cm³ = 1 m*l*

Work out the capacity of the tin in m*l*. *(3 marks)*

SOUP — 11 cm, 7 cm

Radius of base = 7 ÷ 2 = 3.5 cm
Volume = $\pi r^2 h$ = $\pi \times 3.5^2 \times 11$
= 423.3296... cm³

The capacity is 423 m*l* to the nearest whole number.

Now try this (target C)

The diagram shows an oil drum in the shape of a cylinder of height 84 cm and diameter 58 cm.

It is one-quarter full of crude oil.

Calculate the volume of oil in the cylinder.

Give your answer in litres, correct to the nearest litre. *(4 marks)*

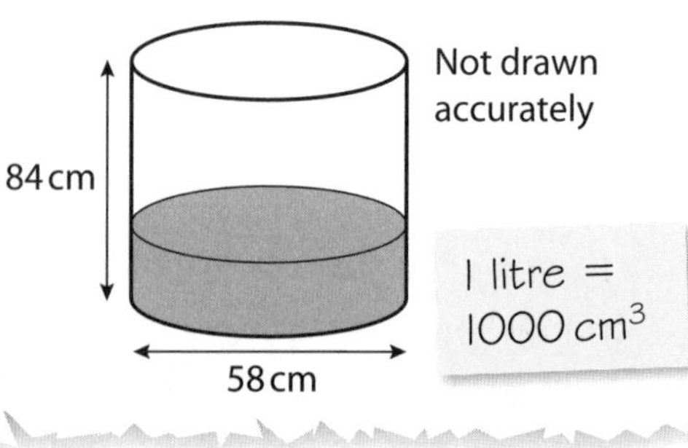

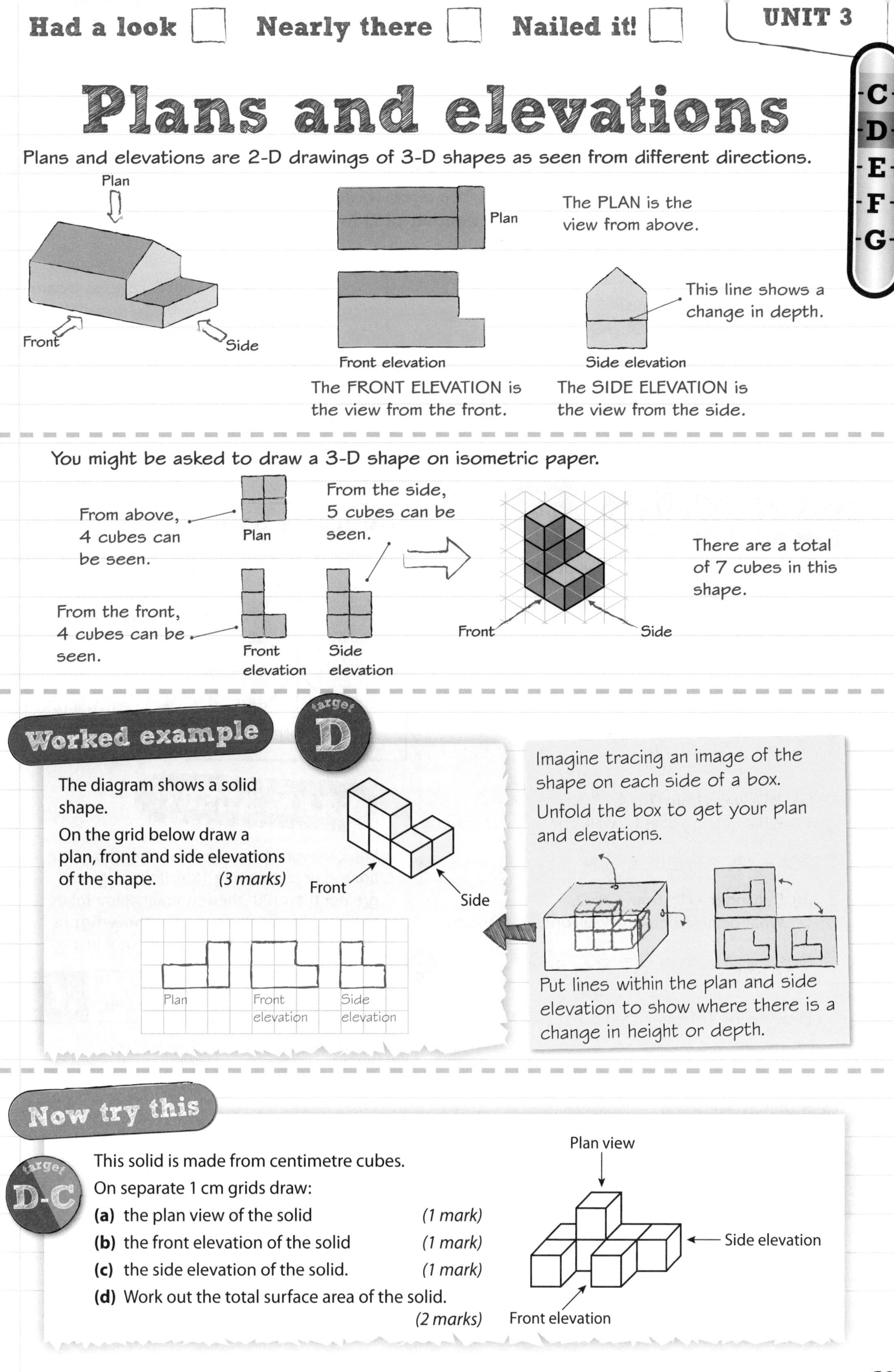

C D E F G

Plans and elevations

Plans and elevations are 2-D drawings of 3-D shapes as seen from different directions.

The FRONT ELEVATION is the view from the front.

The SIDE ELEVATION is the view from the side.

You might be asked to draw a 3-D shape on isometric paper.

There are a total of 7 cubes in this shape.

Worked example

target D

The diagram shows a solid shape.

On the grid below draw a plan, front and side elevations of the shape. *(3 marks)*

Imagine tracing an image of the shape on each side of a box.

Unfold the box to get your plan and elevations.

Put lines within the plan and side elevation to show where there is a change in height or depth.

Now try this

target D-C

This solid is made from centimetre cubes.

On separate 1 cm grids draw:

(a) the plan view of the solid *(1 mark)*

(b) the front elevation of the solid *(1 mark)*

(c) the side elevation of the solid. *(1 mark)*

(d) Work out the total surface area of the solid. *(2 marks)*

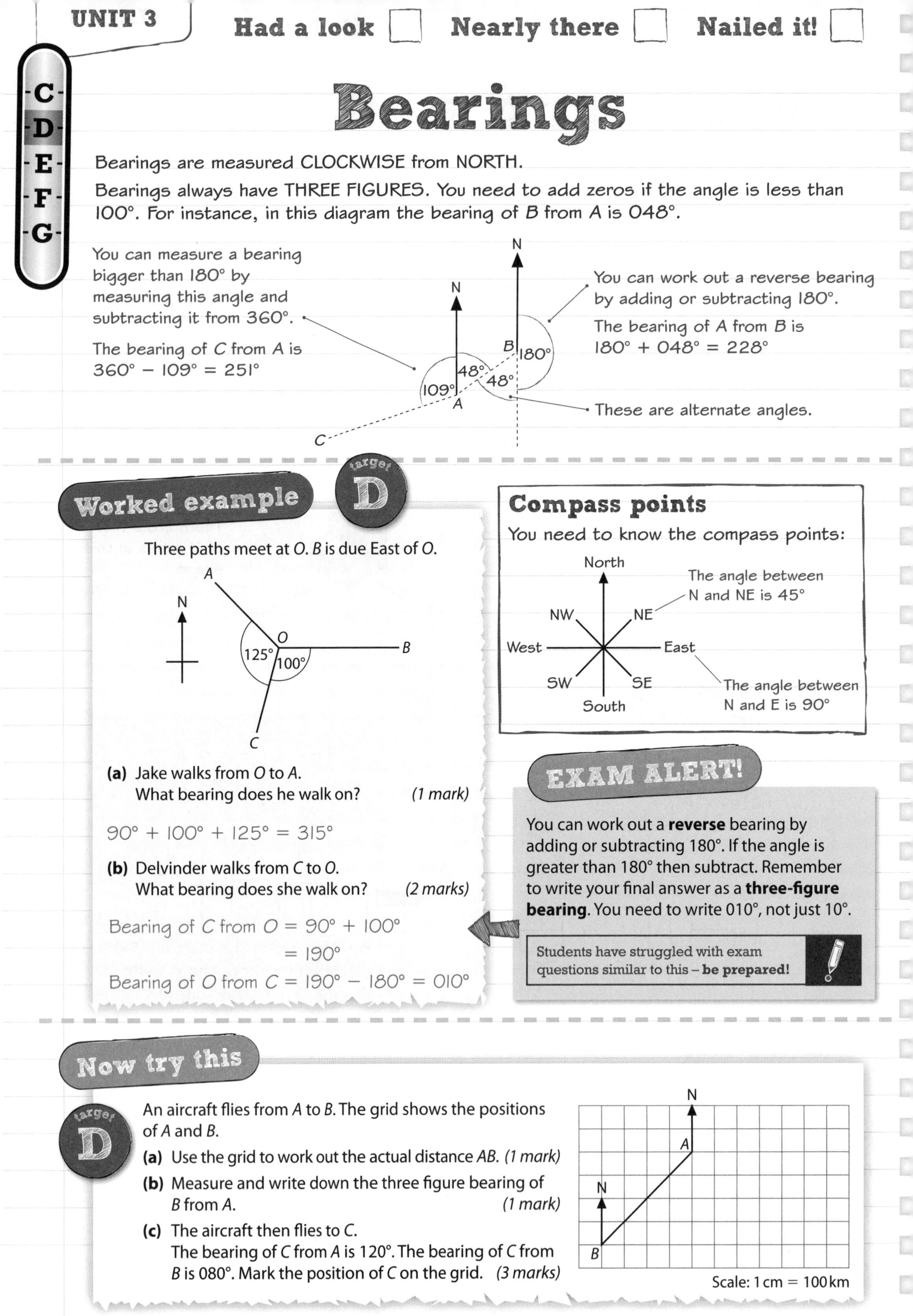

C D E F G

Bearings

Bearings are measured CLOCKWISE from NORTH.

Bearings always have THREE FIGURES. You need to add zeros if the angle is less than 100°. For instance, in this diagram the bearing of B from A is 048°.

You can measure a bearing bigger than 180° by measuring this angle and subtracting it from 360°.

The bearing of C from A is $360° - 109° = 251°$

You can work out a reverse bearing by adding or subtracting 180°.

The bearing of A from B is $180° + 048° = 228°$

These are alternate angles.

Worked example

target D

Three paths meet at O. B is due East of O.

(a) Jake walks from O to A.
What bearing does he walk on? *(1 mark)*

$90° + 100° + 125° = 315°$

(b) Delvinder walks from C to O.
What bearing does she walk on? *(2 marks)*

Bearing of C from $O = 90° + 100°$
$= 190°$

Bearing of O from $C = 190° - 180° = 010°$

Compass points

You need to know the compass points:

EXAM ALERT!

You can work out a **reverse** bearing by adding or subtracting 180°. If the angle is greater than 180° then subtract. Remember to write your final answer as a **three-figure bearing**. You need to write 010°, not just 10°.

Students have struggled with exam questions similar to this – **be prepared!**

Now try this

target D

An aircraft flies from A to B. The grid shows the positions of A and B.

(a) Use the grid to work out the actual distance AB. *(1 mark)*

(b) Measure and write down the three figure bearing of B from A. *(1 mark)*

(c) The aircraft then flies to C.
The bearing of C from A is 120°. The bearing of C from B is 080°. Mark the position of C on the grid. *(3 marks)*

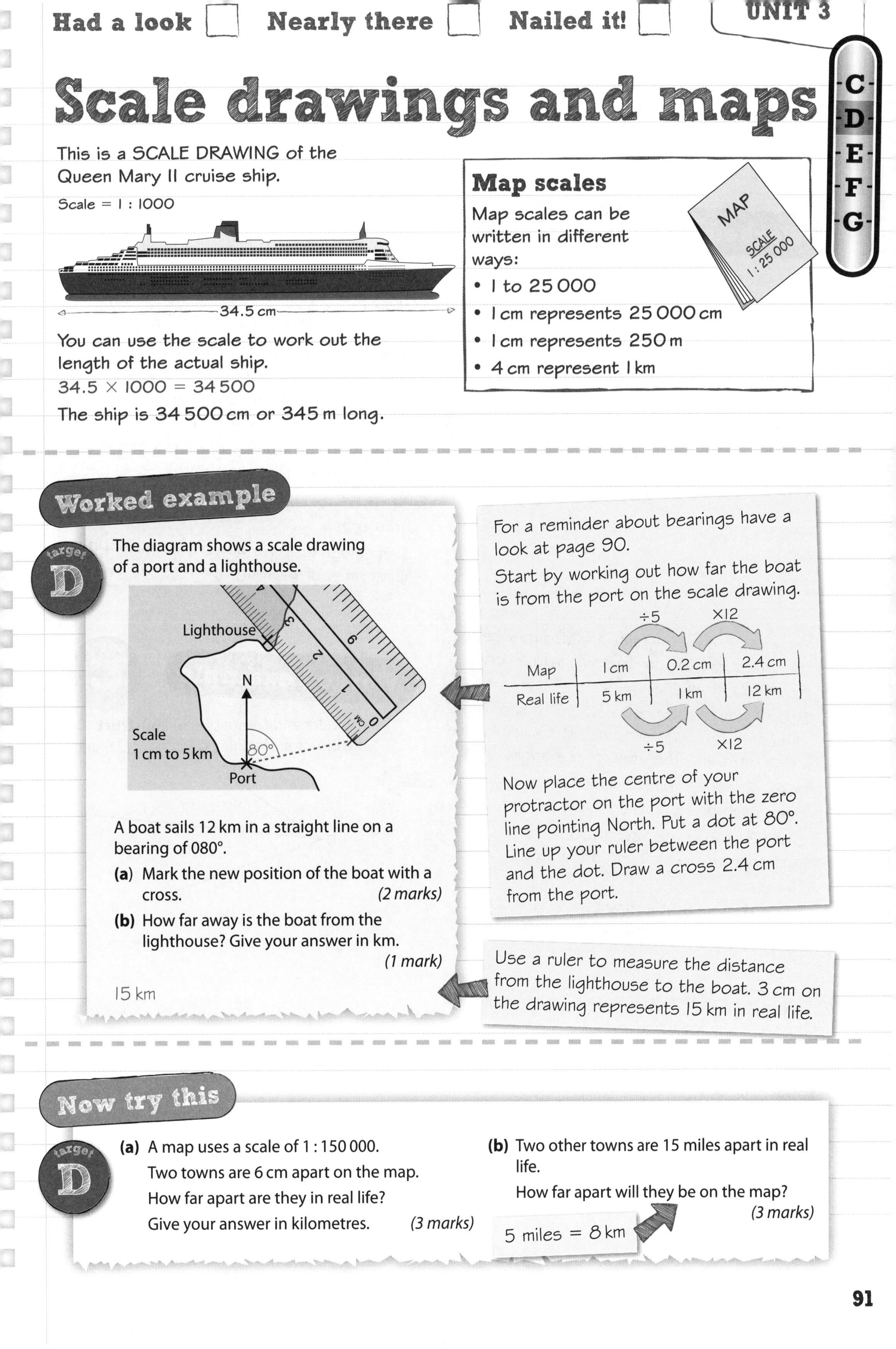

Scale drawings and maps

This is a SCALE DRAWING of the Queen Mary II cruise ship.

Scale = 1 : 1000

You can use the scale to work out the length of the actual ship.

$34.5 \times 1000 = 34\,500$

The ship is 34 500 cm or 345 m long.

Map scales

Map scales can be written in different ways:

- 1 to 25 000
- 1 cm represents 25 000 cm
- 1 cm represents 250 m
- 4 cm represent 1 km

Worked example

Target D

The diagram shows a scale drawing of a port and a lighthouse.

A boat sails 12 km in a straight line on a bearing of 080°.

(a) Mark the new position of the boat with a cross. *(2 marks)*

(b) How far away is the boat from the lighthouse? Give your answer in km. *(1 mark)*

15 km

For a reminder about bearings have a look at page 90.

Start by working out how far the boat is from the port on the scale drawing.

Map	1 cm	0.2 cm	2.4 cm
Real life	5 km	1 km	12 km

Now place the centre of your protractor on the port with the zero line pointing North. Put a dot at 80°. Line up your ruler between the port and the dot. Draw a cross 2.4 cm from the port.

Use a ruler to measure the distance from the lighthouse to the boat. 3 cm on the drawing represents 15 km in real life.

Now try this

Target D

(a) A map uses a scale of 1 : 150 000.
Two towns are 6 cm apart on the map.
How far apart are they in real life?
Give your answer in kilometres. *(3 marks)*

(b) Two other towns are 15 miles apart in real life.
How far apart will they be on the map? *(3 marks)*

5 miles = 8 km

C D E F G

Constructing bisectors

BISECT means 'cut in half'. You need to be able to accurately construct the bisector of an angle and the perpendicular bisector of a line using a ruler and a pair of compasses.

Worked example

target C

Use ruler and compasses to **construct** the perpendicular bisector of the line *AB*. *(2 marks)*

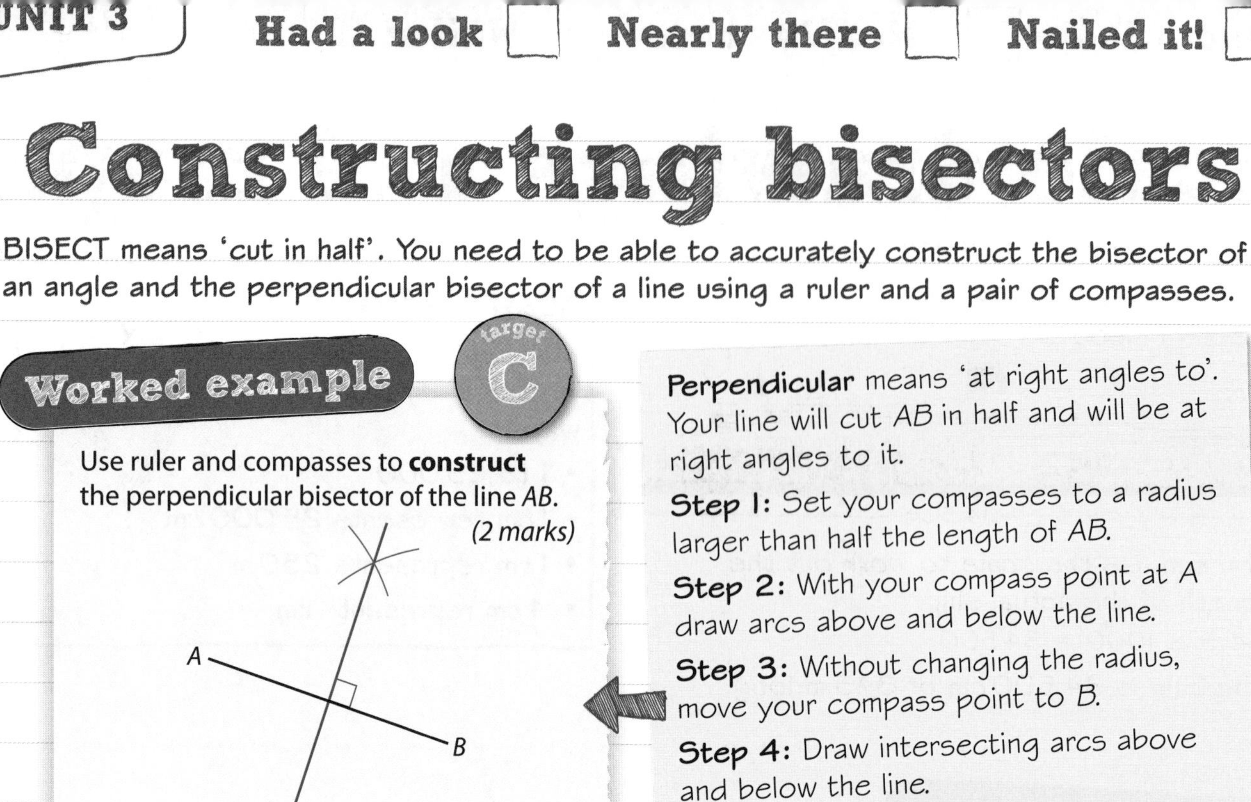

Perpendicular means 'at right angles to'. Your line will cut *AB* in half and will be at right angles to it.

Step 1: Set your compasses to a radius larger than half the length of *AB*.

Step 2: With your compass point at *A* draw arcs above and below the line.

Step 3: Without changing the radius, move your compass point to *B*.

Step 4: Draw intersecting arcs above and below the line.

Step 5: Join up the points where your arcs cross.

Remember to leave **all** your construction lines on your drawing.

You must use only a ruler and compasses in this question. You can't use a protractor to measure the angle.

Step 1: Place the point of your compasses at *Q* and draw two arcs with the same radius which intersect the arms of the angle.

Step 2: Move your compass point to *A* and draw another arc in the middle of the angle.

Step 3: Move your compass point to *B* and draw an intersecting arc with the same radius.

Step 4: Join up the points where the arcs intersect with the point of the angle, *Q*.

Worked example

target C

Use ruler and compasses to **construct** the bisector of angle *PQR*. *(2 marks)*

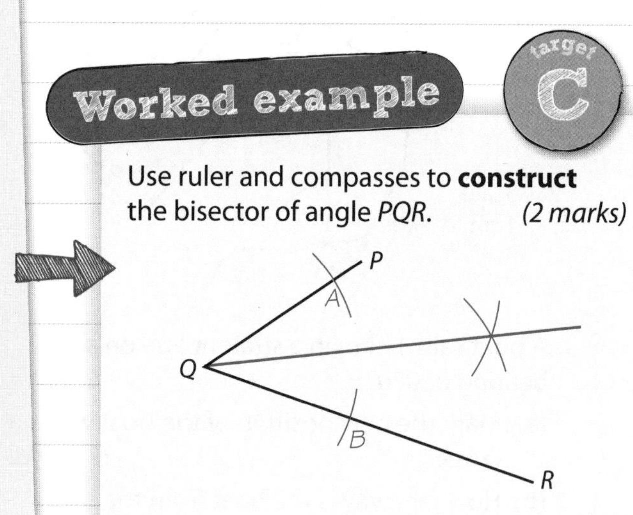

Now try this

target C

Here is triangle *ABC*.

(a) Use a ruler and compasses to construct the perpendicular bisector of side *BC*. *(2 marks)*

(b) Use a ruler and compasses to construct the bisector of angle *A*. *(2 marks)*

(c) Mark the point *X* where your lines intersect. Measure the distance *AX*. *(1 mark)*

Remember to leave all your construction lines on the diagram.

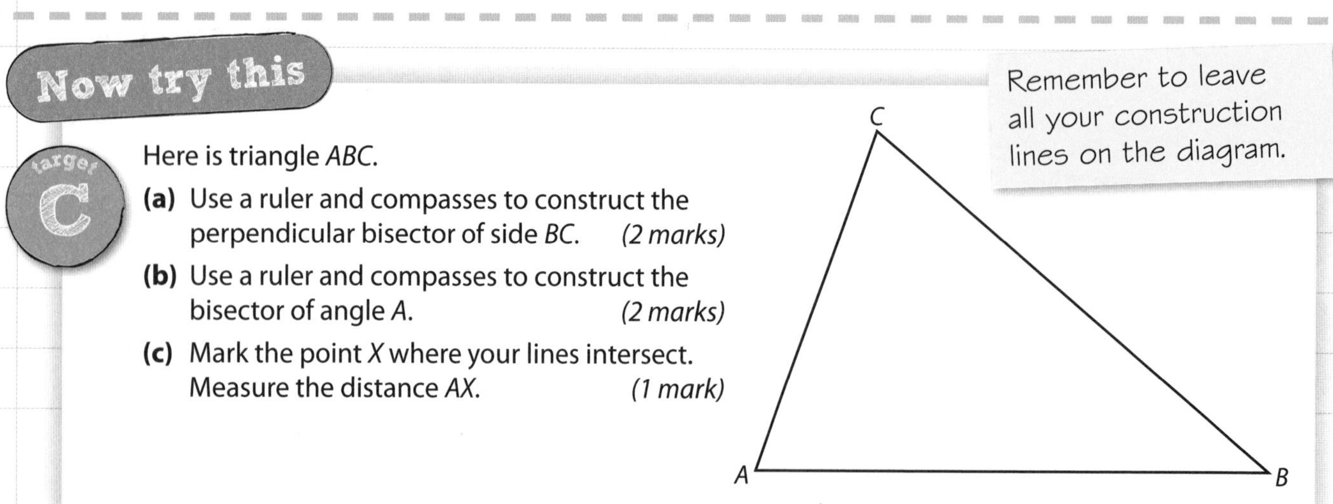

C D E F G

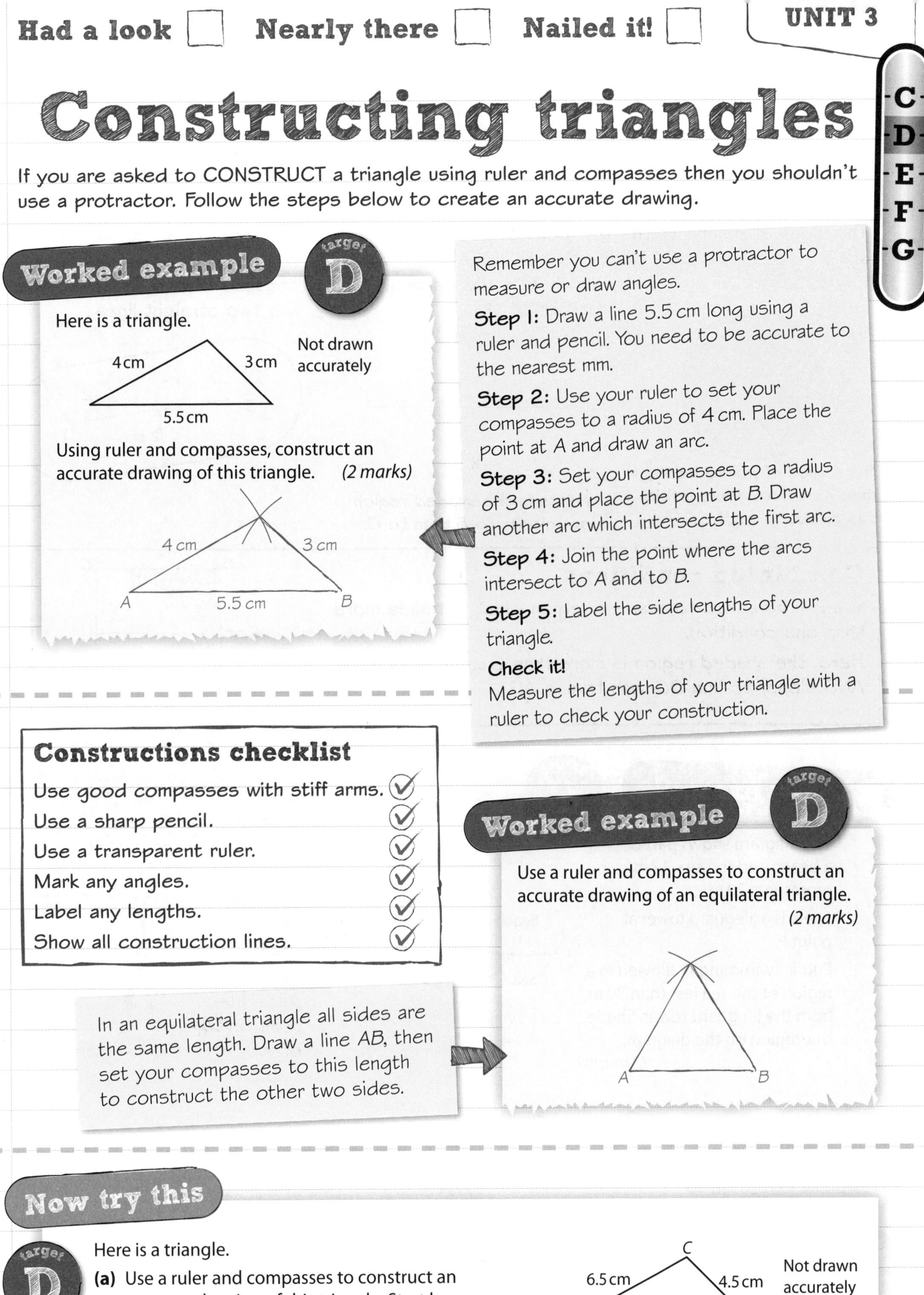

Constructing triangles

If you are asked to CONSTRUCT a triangle using ruler and compasses then you shouldn't use a protractor. Follow the steps below to create an accurate drawing.

Worked example (target D)

Here is a triangle.

Using ruler and compasses, construct an accurate drawing of this triangle. *(2 marks)*

Remember you can't use a protractor to measure or draw angles.

Step 1: Draw a line 5.5 cm long using a ruler and pencil. You need to be accurate to the nearest mm.

Step 2: Use your ruler to set your compasses to a radius of 4 cm. Place the point at A and draw an arc.

Step 3: Set your compasses to a radius of 3 cm and place the point at B. Draw another arc which intersects the first arc.

Step 4: Join the point where the arcs intersect to A and to B.

Step 5: Label the side lengths of your triangle.

Check it!
Measure the lengths of your triangle with a ruler to check your construction.

Constructions checklist

- Use good compasses with stiff arms. ✓
- Use a sharp pencil. ✓
- Use a transparent ruler. ✓
- Mark any angles. ✓
- Label any lengths. ✓
- Show all construction lines. ✓

Worked example (target D)

Use a ruler and compasses to construct an accurate drawing of an equilateral triangle. *(2 marks)*

In an equilateral triangle all sides are the same length. Draw a line AB, then set your compasses to this length to construct the other two sides.

Now try this (target D)

Here is a triangle.

(a) Use a ruler and compasses to construct an accurate drawing of this triangle. Start by drawing the base of the triangle of length 8 cm. *(2 marks)*

(b) In your drawing, measure the size of angle A. *(1 mark)*

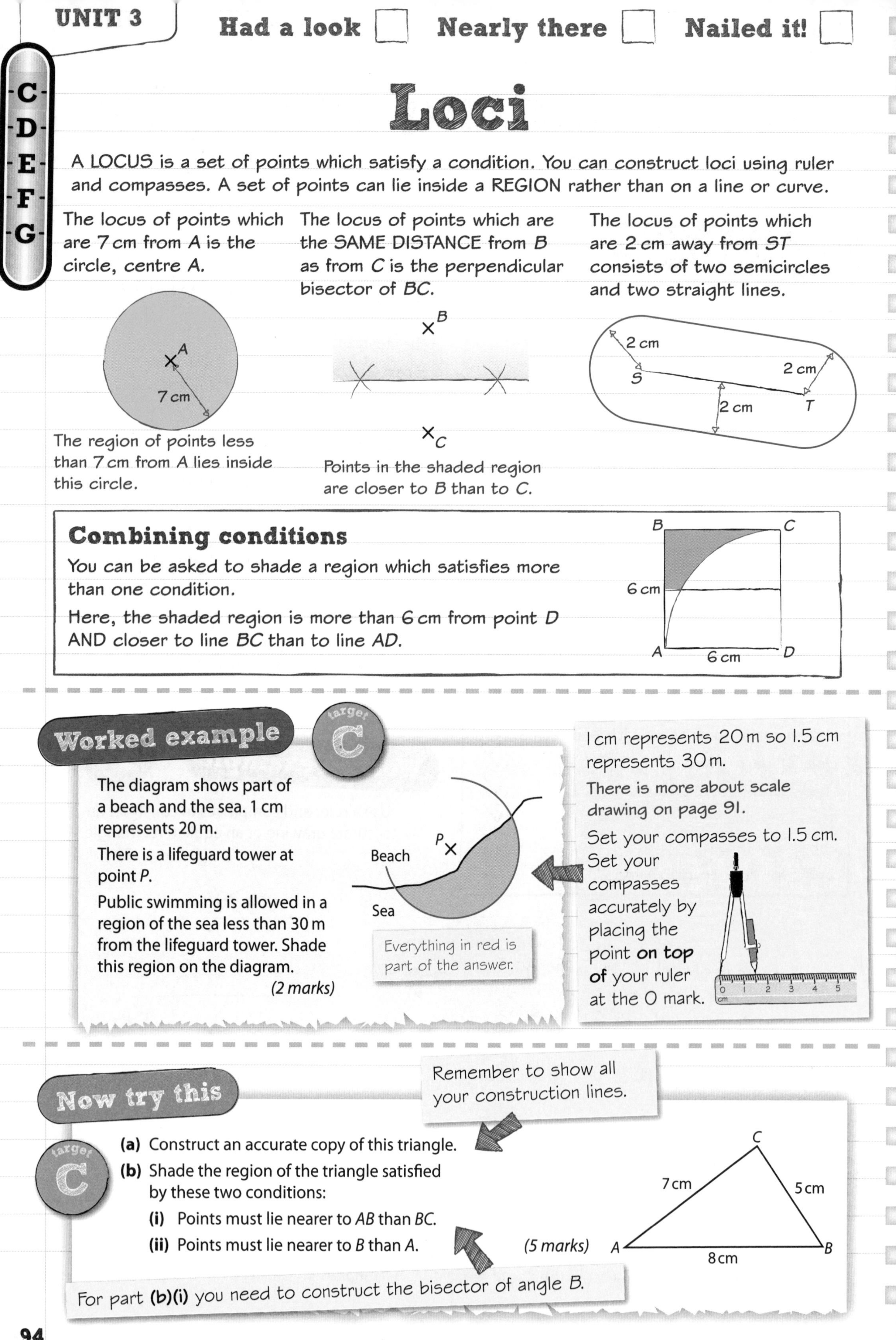

C D E F G

Loci

A LOCUS is a set of points which satisfy a condition. You can construct loci using ruler and compasses. A set of points can lie inside a REGION rather than on a line or curve.

The locus of points which are 7 cm from A is the circle, centre A.

The region of points less than 7 cm from A lies inside this circle.

The locus of points which are the SAME DISTANCE from B as from C is the perpendicular bisector of BC.

Points in the shaded region are closer to B than to C.

The locus of points which are 2 cm away from ST consists of two semicircles and two straight lines.

Combining conditions

You can be asked to shade a region which satisfies more than one condition.

Here, the shaded region is more than 6 cm from point D AND closer to line BC than to line AD.

Worked example

target C

The diagram shows part of a beach and the sea. 1 cm represents 20 m.

There is a lifeguard tower at point P.

Public swimming is allowed in a region of the sea less than 30 m from the lifeguard tower. Shade this region on the diagram.

(2 marks)

Everything in red is part of the answer.

1 cm represents 20 m so 1.5 cm represents 30 m.

There is more about scale drawing on page 91.

Set your compasses to 1.5 cm.

Set your compasses accurately by placing the point **on top of** your ruler at the 0 mark.

Now try this

target C

Remember to show all your construction lines.

(a) Construct an accurate copy of this triangle.

(b) Shade the region of the triangle satisfied by these two conditions:

(i) Points must lie nearer to AB than BC.

(ii) Points must lie nearer to B than A.

(5 marks)

For part **(b)(i)** you need to construct the bisector of angle B.

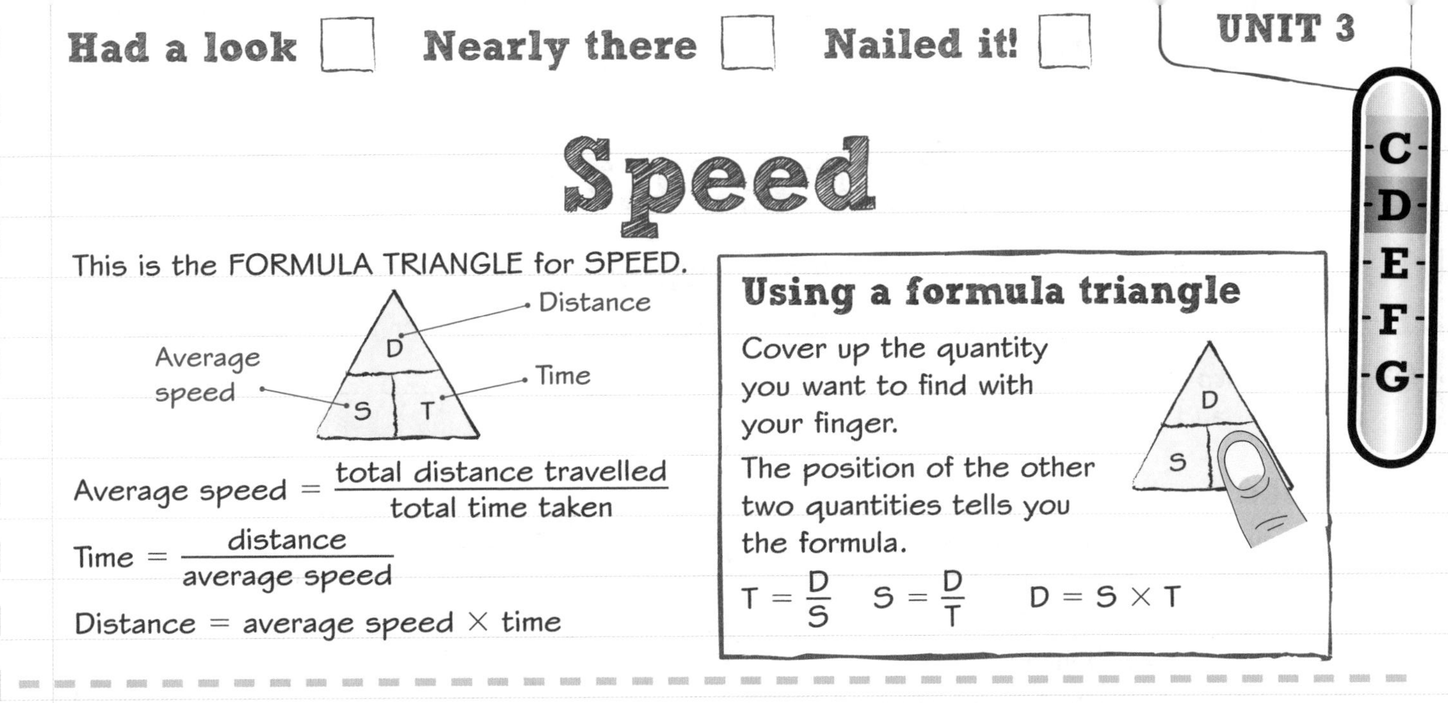

C D E F G

Speed

This is the FORMULA TRIANGLE for SPEED.

Average speed = $\frac{\text{total distance travelled}}{\text{total time taken}}$

Time = $\frac{\text{distance}}{\text{average speed}}$

Distance = average speed × time

Using a formula triangle

Cover up the quantity you want to find with your finger.

The position of the other two quantities tells you the formula.

$T = \frac{D}{S}$ $S = \frac{D}{T}$ $D = S \times T$

Units

The most common units of speed are

- metres per second: m/s
- kilometres per hour: km/h
- miles per hour: mph

The units in your answer will depend on the units you use in the formula.

When distance is measured in km and time is measured in hours, speed will be measured in km/h.

Minutes and hours

For questions on speed, you need to be able to convert between minutes and hours.

Hours	Minutes
$\frac{1}{2}$	30
$\frac{3}{4}$	45
$2\frac{1}{4}$	135

You can also write this as '2 hours and 15 minutes'. 60 + 60 + 15 = 135

Worked example

target D

A plane travels at a constant speed of 600 km/h for 45 minutes.
How far has it travelled? *(3 marks)*

D / S T

45 minutes = $\frac{45}{60}$ hour = $\frac{3}{4}$ hour

$D = S \times T$

$= 600 \times \frac{3}{4} = \frac{600 \times 3}{4} = \frac{1800}{4} = 450$

The plane has travelled 450 km.

Speed questions

- Draw a formula triangle. ✓
- Make sure the units match. ✓
- Give units with your answer. ✓

EXAM ALERT!

Make sure that the units match. Speed is given in km/h, so convert the time into hours by dividing by 60. The units of distance will be km.

Students have struggled with exam questions similar to this – **be prepared!**

Now try this

target D

1 Bradley cycled 170 km at an average speed of 40 kilometres per hour.
How long did it take him?
Give your answer in hours and minutes. *(3 marks)*

2 Simon drives from Newcastle to Oxford.
His average speed is 56 mph.
The journey takes 4 hours 45 minutes.
How far did he drive? *(3 marks)*

C D E F G

Measures

Most of the units of measurement used in the UK are METRIC units like litres, kilograms and kilometres. You can convert these into IMPERIAL units like gallons, pounds and miles.

Speed

To convert between measures of speed using metric units you need to convert one unit first then the other.

To convert 72 km/h into m/s:

72 km/h → 72 × 1000 = 72 000 m/h

72 000 m/h → 72 000 ÷ 3600 = 20 m/s

1 hour = 60 × 60 = 3600 seconds

Metric and imperial

You need to remember these conversions for your exam.

Metric unit	Imperial unit
1 kg	2.2 pounds (lb)
1 litre (l)	$1\frac{3}{4}$ pints
4.5 litres	1 gallon
8 km	5 miles
2.5 cm	1 inch

When converting between different imperial units you will be GIVEN the conversions.

For a reminder about metric units have a look at page 73.

Worked example

target D

You are given that 14 lb = 1 stone.
Alicia's dog weighs 3 stone.
How many kg does Alicia's dog weigh? *(2 marks)*

3 × 14 = 42
42 ÷ 2.2 = 19.0909…
Alicia's dog weighs 19 kg to the nearest kg.

You will be given conversions between different imperial units, like pounds and stone.
You know that 1 kg = 2.2 lb. So start by converting 3 stone into lb.
Then divide by 2.2 to convert lb into kg.

Worked example

target D

Heather drives along a motorway at an average speed of 60 mph.
Work out how long it takes her to drive 120 km. *(3 marks)*

8 km = 5 miles
60 ÷ 5 = 12
8 × 12 = 96
60 mph = 96 km/h

$$\text{Time} = \frac{\text{distance}}{\text{average speed}} = \frac{120}{96} = 1\tfrac{1}{4} \text{ hours}$$

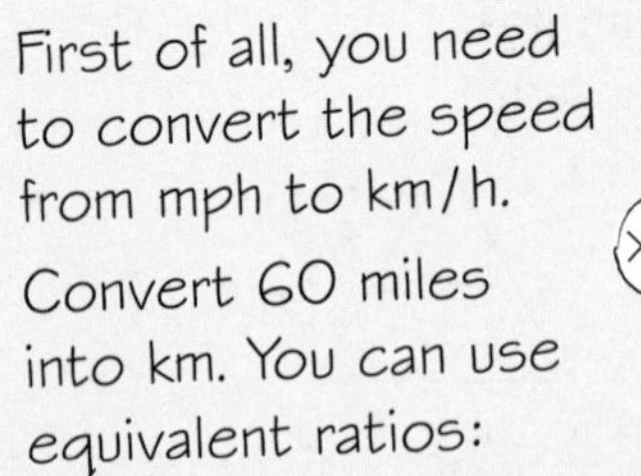

First of all, you need to convert the speed from mph to km/h.
Convert 60 miles into km. You can use equivalent ratios:

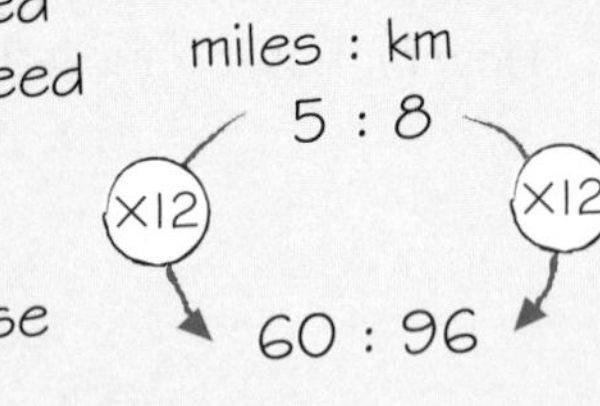

Kilometres are smaller than miles so the speed in km/h should be a larger number than the speed in mph.
Then use the formula triangle for speed (see page 95) to work out the answer.

Now try this

target D

1 In France, the speed limit on motorways, when it is dry, is 130 km/h.
What speed is this in mph? *(2 marks)*

2 Ian's car has a petrol tank that holds 12 gallons.
He fills the tank, from empty, with petrol that costs £1.32 per litre.
How much should he pay for this petrol? *(3 marks)*

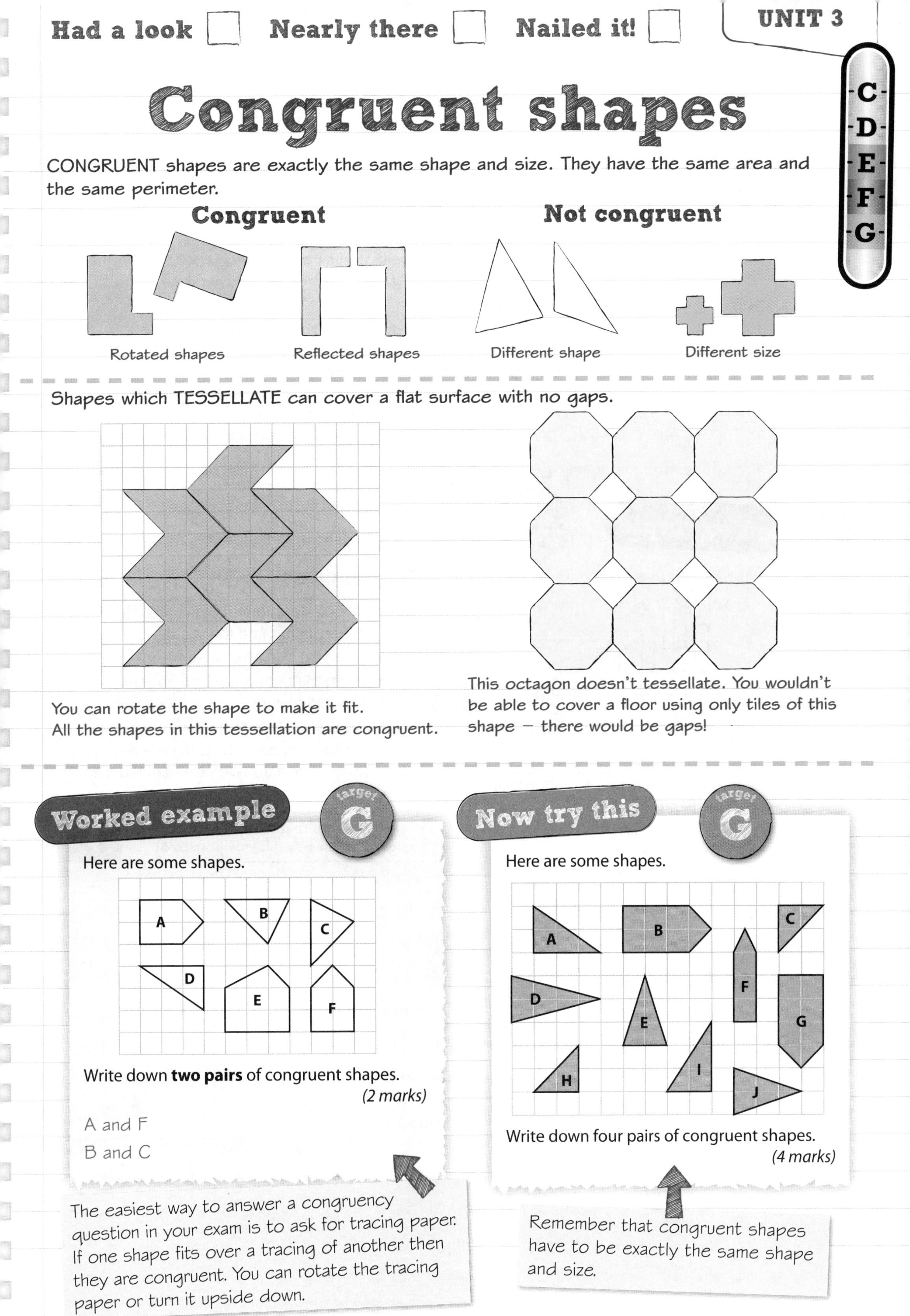

Congruent shapes

CONGRUENT shapes are exactly the same shape and size. They have the same area and the same perimeter.

Congruent

Rotated shapes

Reflected shapes

Not congruent

Different shape

Different size

Shapes which TESSELLATE can cover a flat surface with no gaps.

You can rotate the shape to make it fit.
All the shapes in this tessellation are congruent.

This octagon doesn't tessellate. You wouldn't be able to cover a floor using only tiles of this shape – there would be gaps!

Worked example

target G

Here are some shapes.

A B C D E F

Write down **two pairs** of congruent shapes.

(2 marks)

A and F
B and C

The easiest way to answer a congruency question in your exam is to ask for tracing paper. If one shape fits over a tracing of another then they are congruent. You can rotate the tracing paper or turn it upside down.

Now try this

target G

Here are some shapes.

A B C D E F G H I J

Write down four pairs of congruent shapes.

(4 marks)

Remember that congruent shapes have to be exactly the same shape and size.

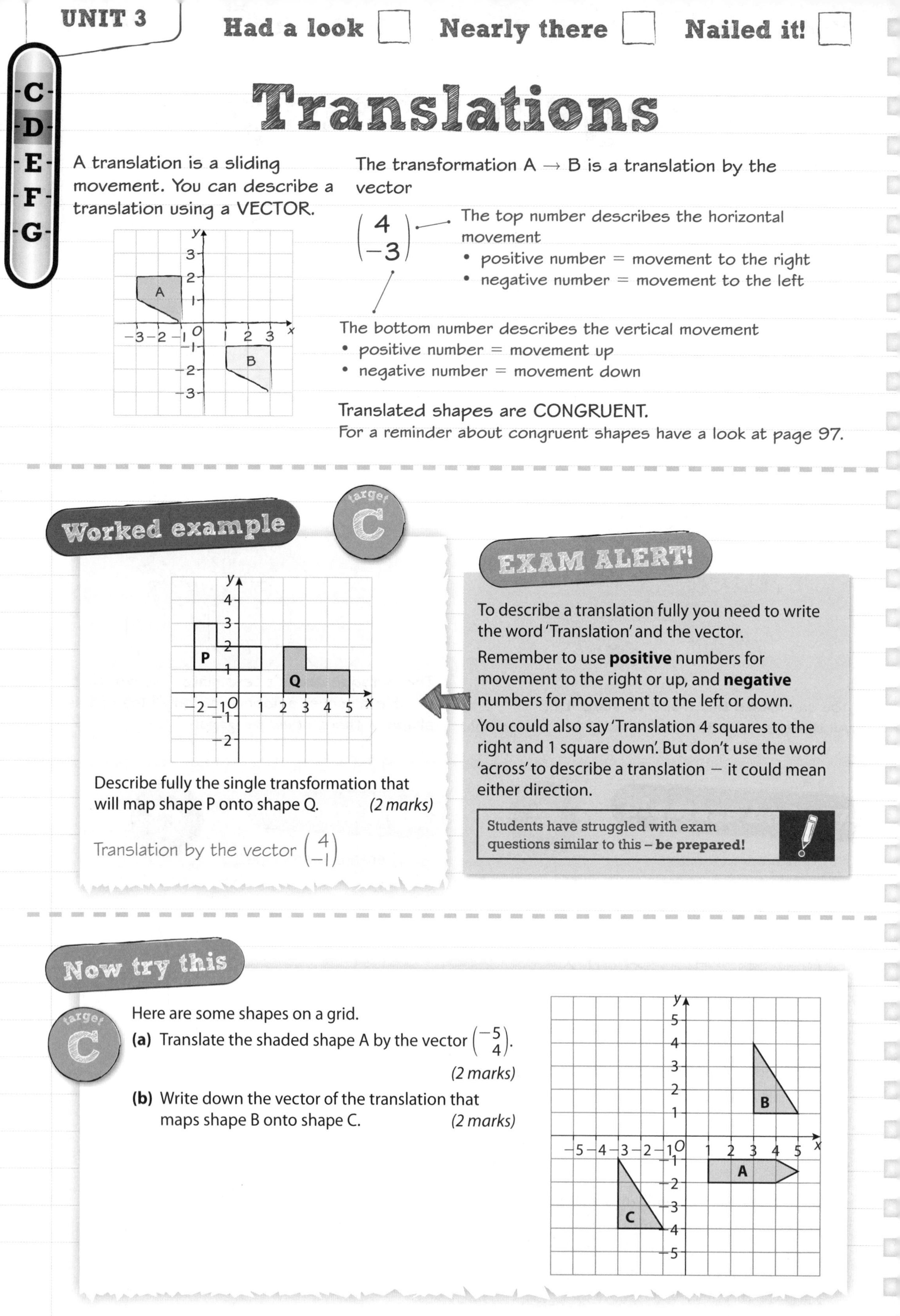

C D E F G

Translations

A translation is a sliding movement. You can describe a translation using a VECTOR.

The transformation A → B is a translation by the vector

$$\begin{pmatrix} 4 \\ -3 \end{pmatrix}$$

The top number describes the horizontal movement
- positive number = movement to the right
- negative number = movement to the left

The bottom number describes the vertical movement
- positive number = movement up
- negative number = movement down

Translated shapes are CONGRUENT.

For a reminder about congruent shapes have a look at page 97.

Worked example

target C

Describe fully the single transformation that will map shape P onto shape Q. *(2 marks)*

Translation by the vector $\begin{pmatrix} 4 \\ -1 \end{pmatrix}$

EXAM ALERT!

To describe a translation fully you need to write the word 'Translation' and the vector.

Remember to use **positive** numbers for movement to the right or up, and **negative** numbers for movement to the left or down.

You could also say 'Translation 4 squares to the right and 1 square down'. But don't use the word 'across' to describe a translation – it could mean either direction.

Students have struggled with exam questions similar to this – **be prepared!**

Now try this

target C

Here are some shapes on a grid.

(a) Translate the shaded shape A by the vector $\begin{pmatrix} -5 \\ 4 \end{pmatrix}$. *(2 marks)*

(b) Write down the vector of the translation that maps shape B onto shape C. *(2 marks)*

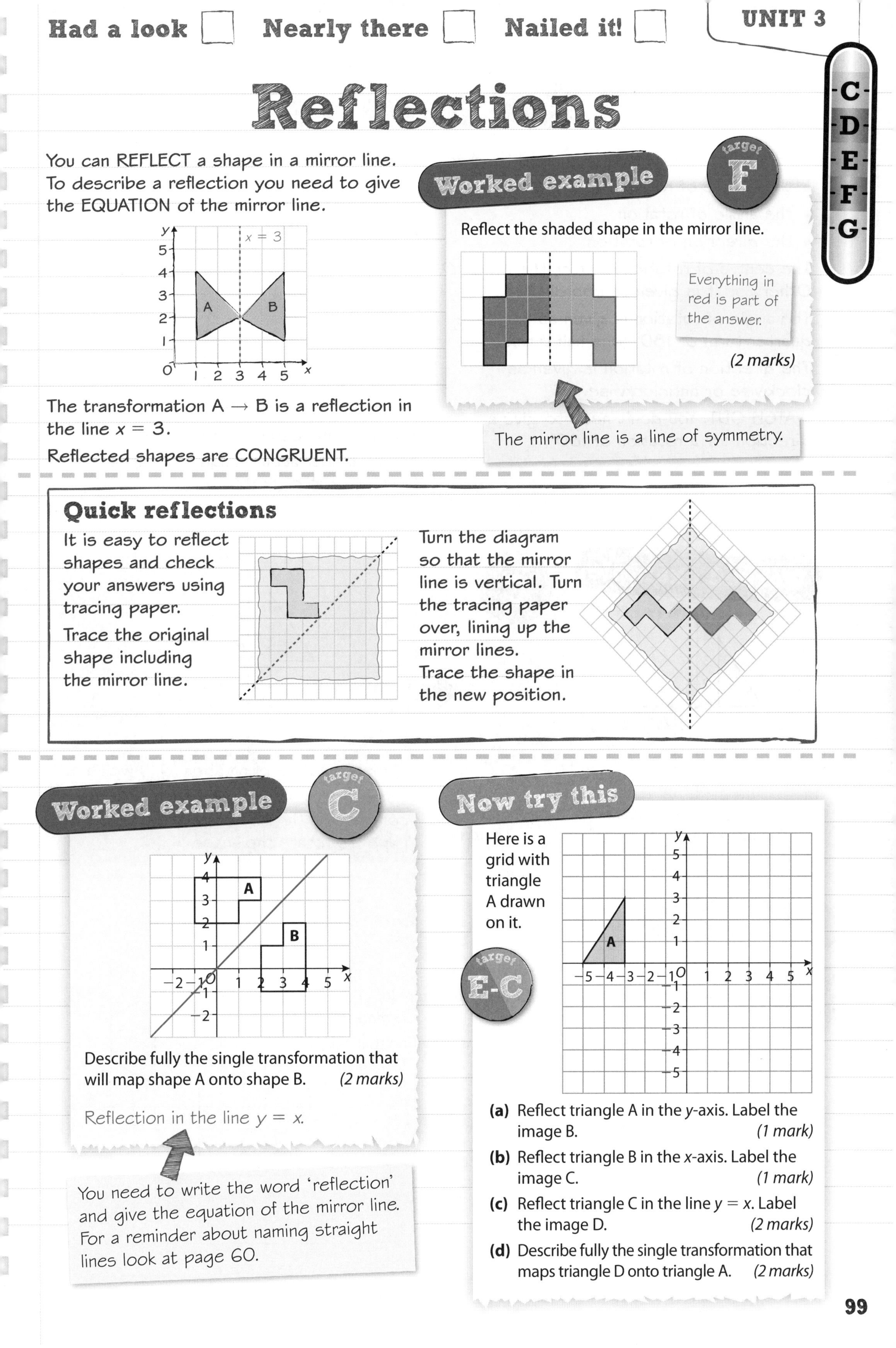

C D E F G

Reflections

You can REFLECT a shape in a mirror line. To describe a reflection you need to give the EQUATION of the mirror line.

The transformation A → B is a reflection in the line $x = 3$.

Reflected shapes are CONGRUENT.

Worked example

target F

Reflect the shaded shape in the mirror line.

Everything in red is part of the answer.

(2 marks)

The mirror line is a line of symmetry.

Quick reflections

It is easy to reflect shapes and check your answers using tracing paper.

Trace the original shape including the mirror line.

Turn the diagram so that the mirror line is vertical. Turn the tracing paper over, lining up the mirror lines.

Trace the shape in the new position.

Worked example

target C

Describe fully the single transformation that will map shape A onto shape B. *(2 marks)*

Reflection in the line $y = x$.

You need to write the word 'reflection' and give the equation of the mirror line. For a reminder about naming straight lines look at page 60.

Now try this

target E-C

Here is a grid with triangle A drawn on it.

(a) Reflect triangle A in the y-axis. Label the image B. *(1 mark)*

(b) Reflect triangle B in the x-axis. Label the image C. *(1 mark)*

(c) Reflect triangle C in the line $y = x$. Label the image D. *(2 marks)*

(d) Describe fully the single transformation that maps triangle D onto triangle A. *(2 marks)*

C D E F G

Rotations

To describe a ROTATION you need to give
- the centre of rotation
- the angle of rotation
- the direction of rotation.

The centre of rotation is often the origin O. Otherwise it is given as coordinates.

The angle of rotation is given as 90° (one quarter turn) or 180° (one half turn).

The direction of rotation is given as clockwise or anticlockwise.

WATCH OUT! You don't need to give a direction for a rotation of 180°.

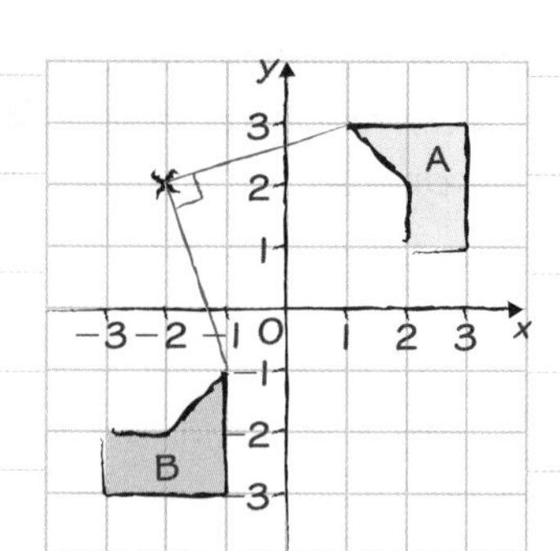

A to B: Rotation 90° clockwise about the point (−2, 2).

You could also say 'a quarter turn clockwise about (−2, 2)'.

You are allowed to ask for tracing paper in the exam. This makes it really easy to rotate shapes and check your answers.

Rotated shapes are CONGRUENT.

For a reminder about congruent shapes have a look at page 97.

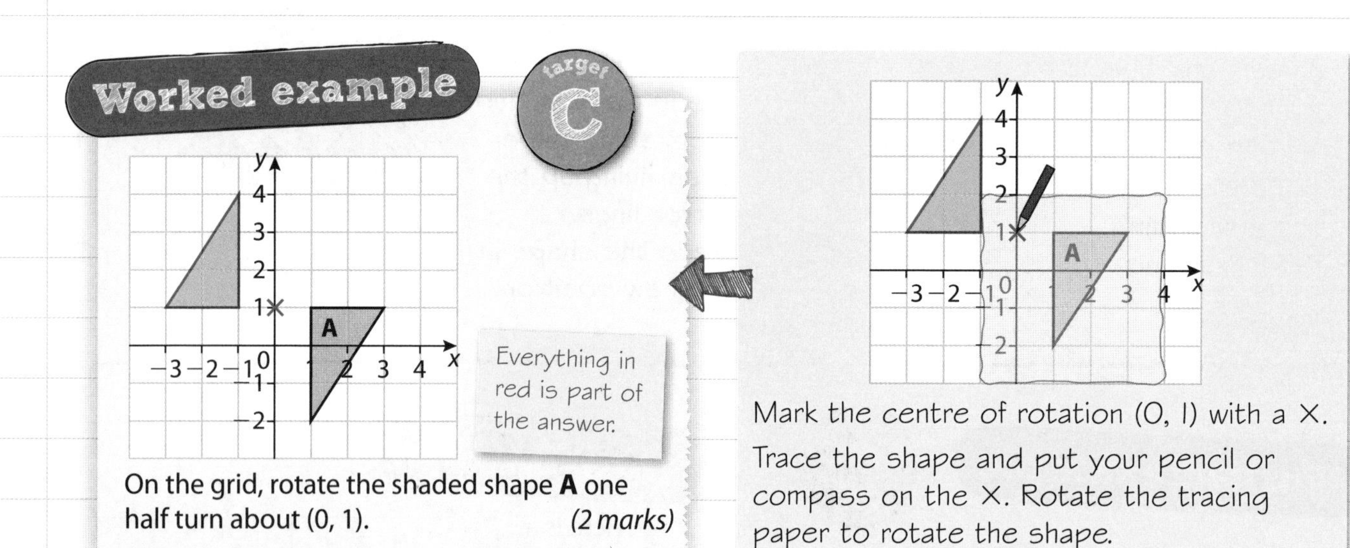

On the grid, rotate the shaded shape **A** one half turn about (0, 1). *(2 marks)*

Mark the centre of rotation (0, 1) with a ×. Trace the shape and put your pencil or compass on the ×. Rotate the tracing paper to rotate the shape.

Now try this

target C

Here is a grid with shapes A and B drawn on it.

(a) Rotate shape B, 90° anticlockwise about the point (0, −1). Label the image C. *(3 marks)*

(b) Describe fully the single rotational transformation that maps shape A onto shape B. *(3 marks)*

To describe a rotation you need to write 'rotation' and give the angle of turn, the direction of the turn and the centre of rotation.

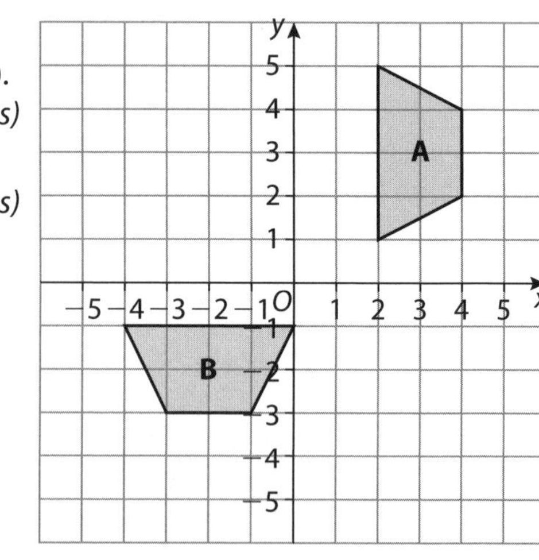

Enlargements

To describe an ENLARGEMENT you need to give the scale factor and the centre of enlargement.

The SCALE FACTOR of an enlargement tells you how much each length is multiplied by.

A to B: Each point on B is twice as far from C as the corresponding point on A.

The transformation A → B is an enlargement with scale factor 2, centre (1, 4).

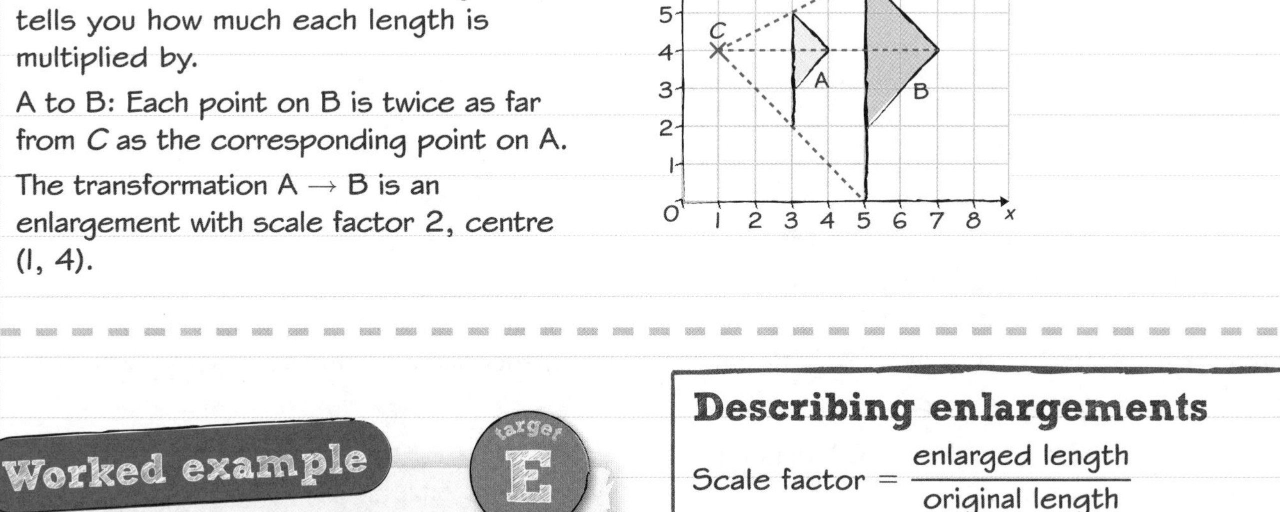

Worked example

target E

On the grid, enlarge the shape with a scale factor of 2. *(2 marks)*

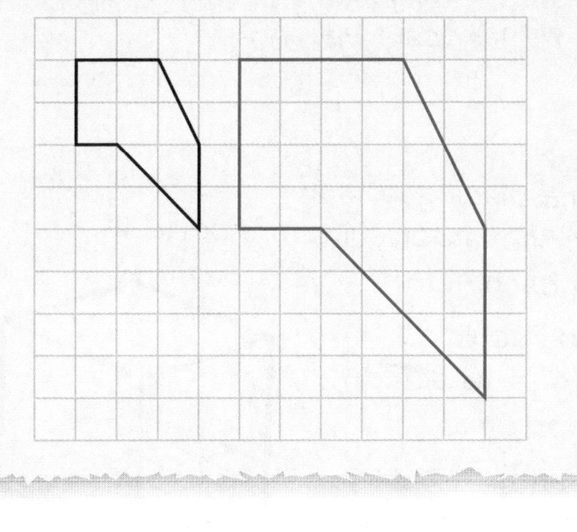

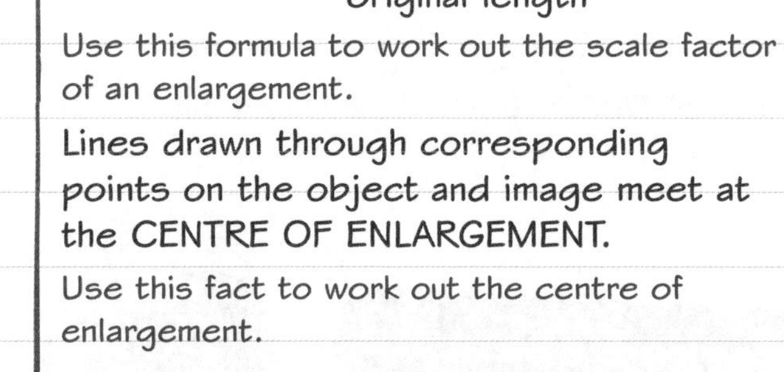

Describing enlargements

$$\text{Scale factor} = \frac{\text{enlarged length}}{\text{original length}}$$

Use this formula to work out the scale factor of an enlargement.

Lines drawn through corresponding points on the object and image meet at the CENTRE OF ENLARGEMENT.

Use this fact to work out the centre of enlargement.

If no centre of enlargement is given, you can draw your new shape anywhere on the grid. Just make sure that every length on the enlarged shape is **2 times** the corresponding length on the original shape.

Now try this

Here is a grid with two shapes A and B drawn on it.

Shape B is an enlargement of shape A.

(a) Write down the scale factor of the enlargement. *(1 mark)*

(b) Work out the position of the centre of enlargement, label it X. *(2 marks)*

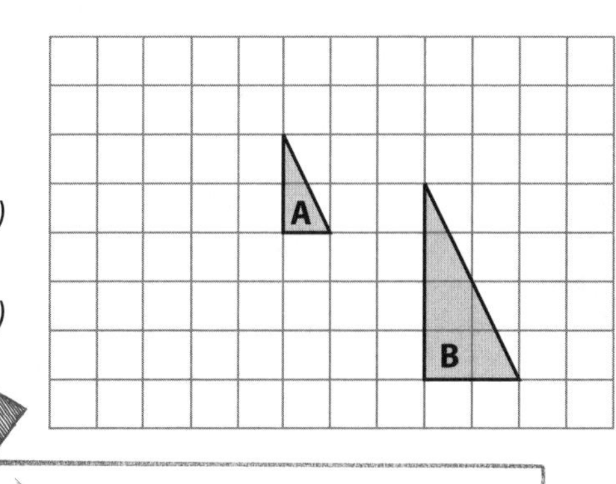

Draw lines through corresponding points on A and B. The centre of rotation will be the point where these lines cross.

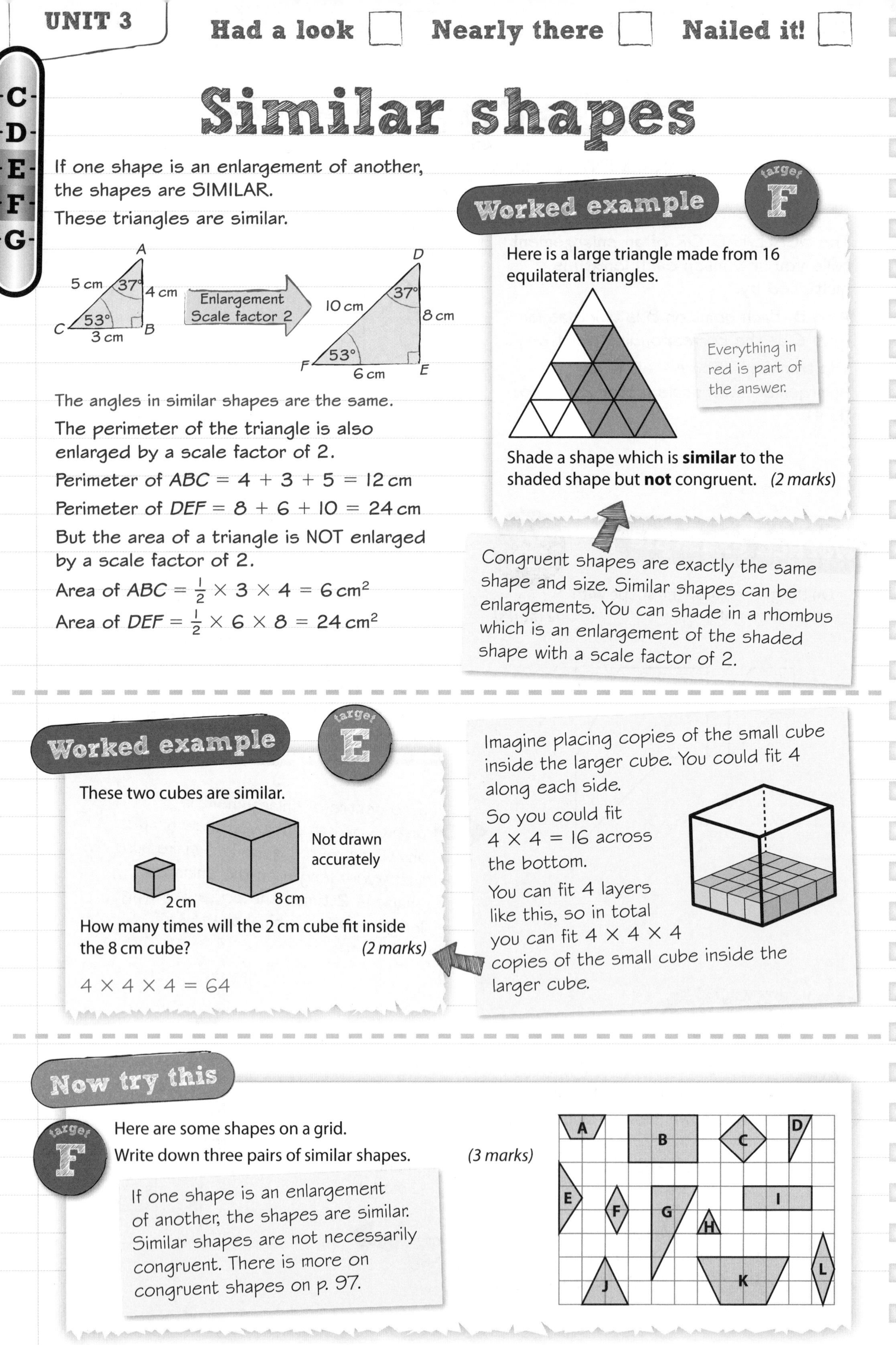

C D E F G

Similar shapes

If one shape is an enlargement of another, the shapes are SIMILAR.

These triangles are similar.

The angles in similar shapes are the same.

The perimeter of the triangle is also enlarged by a scale factor of 2.

Perimeter of $ABC = 4 + 3 + 5 = 12$ cm

Perimeter of $DEF = 8 + 6 + 10 = 24$ cm

But the area of a triangle is NOT enlarged by a scale factor of 2.

Area of $ABC = \frac{1}{2} \times 3 \times 4 = 6\text{ cm}^2$

Area of $DEF = \frac{1}{2} \times 6 \times 8 = 24\text{ cm}^2$

Worked example (target F)

Here is a large triangle made from 16 equilateral triangles.

Everything in red is part of the answer.

Shade a shape which is **similar** to the shaded shape but **not** congruent. *(2 marks)*

Congruent shapes are exactly the same shape and size. Similar shapes can be enlargements. You can shade in a rhombus which is an enlargement of the shaded shape with a scale factor of 2.

Worked example (target E)

These two cubes are similar.

How many times will the 2 cm cube fit inside the 8 cm cube? *(2 marks)*

$4 \times 4 \times 4 = 64$

Imagine placing copies of the small cube inside the larger cube. You could fit 4 along each side.

So you could fit $4 \times 4 = 16$ across the bottom.

You can fit 4 layers like this, so in total you can fit $4 \times 4 \times 4$ copies of the small cube inside the larger cube.

Now try this (target F)

Here are some shapes on a grid.

Write down three pairs of similar shapes. *(3 marks)*

If one shape is an enlargement of another, the shapes are similar. Similar shapes are not necessarily congruent. There is more on congruent shapes on p. 97.

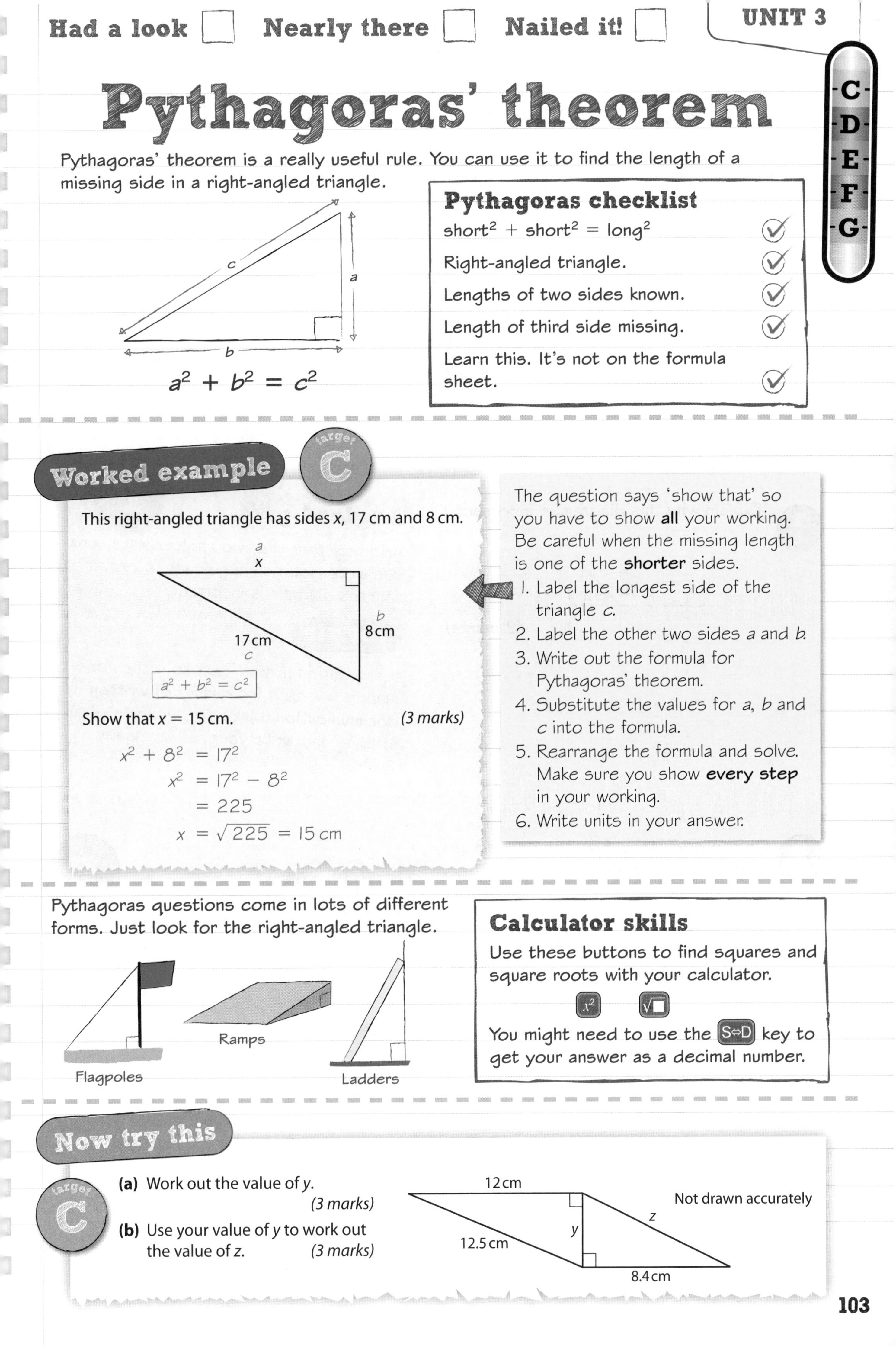

Pythagoras' theorem

Pythagoras' theorem is a really useful rule. You can use it to find the length of a missing side in a right-angled triangle.

$$a^2 + b^2 = c^2$$

Pythagoras checklist

- short^2 + short^2 = long^2 ✓
- Right-angled triangle. ✓
- Lengths of two sides known. ✓
- Length of third side missing. ✓
- Learn this. It's not on the formula sheet. ✓

Worked example (target C)

This right-angled triangle has sides x, 17 cm and 8 cm.

Show that x = 15 cm. *(3 marks)*

$$x^2 + 8^2 = 17^2$$
$$x^2 = 17^2 - 8^2$$
$$= 225$$
$$x = \sqrt{225} = 15\text{ cm}$$

The question says 'show that' so you have to show **all** your working. Be careful when the missing length is one of the **shorter** sides.

1. Label the longest side of the triangle c.
2. Label the other two sides a and b.
3. Write out the formula for Pythagoras' theorem.
4. Substitute the values for a, b and c into the formula.
5. Rearrange the formula and solve. Make sure you show **every step** in your working.
6. Write units in your answer.

Pythagoras questions come in lots of different forms. Just look for the right-angled triangle.

Calculator skills

Use these buttons to find squares and square roots with your calculator.

You might need to use the S⇔D key to get your answer as a decimal number.

Now try this (target C)

(a) Work out the value of y. *(3 marks)*

(b) Use your value of y to work out the value of z. *(3 marks)*

Problem-solving practice 1

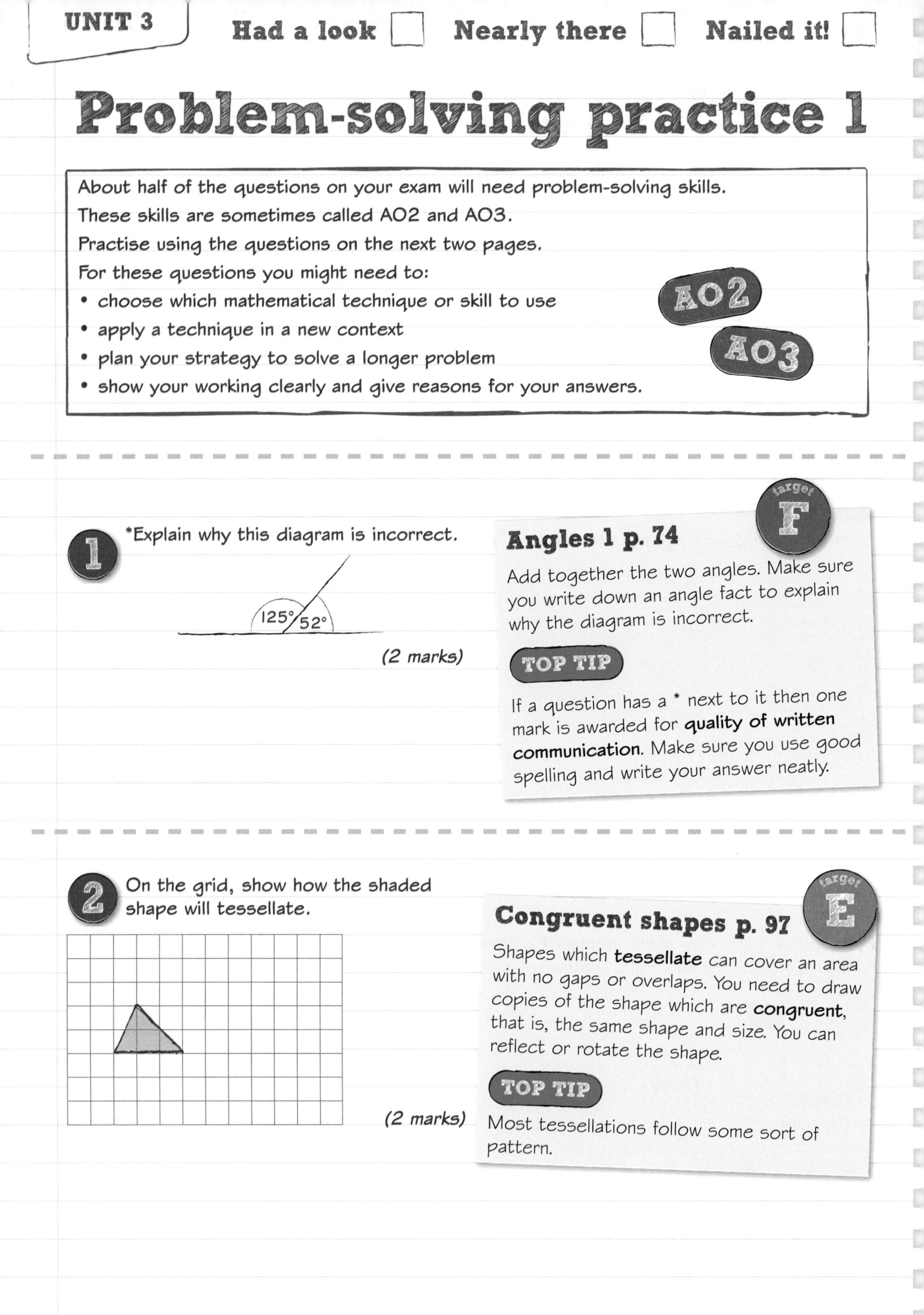

About half of the questions on your exam will need problem-solving skills.
These skills are sometimes called AO2 and AO3.
Practise using the questions on the next two pages.
For these questions you might need to:

- choose which mathematical technique or skill to use
- apply a technique in a new context
- plan your strategy to solve a longer problem
- show your working clearly and give reasons for your answers.

AO2 AO3

1 *Explain why this diagram is incorrect.

(2 marks)

Angles 1 p. 74 — target F

Add together the two angles. Make sure you write down an angle fact to explain why the diagram is incorrect.

TOP TIP

If a question has a * next to it then one mark is awarded for **quality of written communication**. Make sure you use good spelling and write your answer neatly.

2 On the grid, show how the shaded shape will tessellate.

(2 marks)

Congruent shapes p. 97 — target E

Shapes which **tessellate** can cover an area with no gaps or overlaps. You need to draw copies of the shape which are **congruent**, that is, the same shape and size. You can reflect or rotate the shape.

TOP TIP

Most tessellations follow some sort of pattern.

Problem-solving practice 2

3 This diagram shows a quadrilateral.

3y 5y 2y

Work out the value of y. *(3 marks)*

target D

Angles 2 p. 75

Using algebra p. 57

Start by writing an expression for the sum of the angles in the quadrilateral. Use angle facts to write down what this sum is equal to. This gives you an equation you can solve to find the value of y.

TOP TIP

If you know what unknown angles should add up to, then you can write an equation and solve it to find the value of the unknown.

4 A buoy is 6 km from a ship on a bearing of 290°.

A lighthouse is 8 km east of the ship.

Work out the distance between the buoy and the lighthouse. *(3 marks)*

target D

Bearings p. 90

Scale drawings and maps p. 91

You'll need to draw a scale diagram to solve this question. A good scale to use would be 1 cm = 2 km. Always use a ruler and a sharp pencil to draw any lines and **don't** rub out any construction lines or working.

TOP TIP

Bearings less than 180° are to the **right** of North. Bearings between 180° and 360° are to the **left**. Bearings always have three figures.

5

y −6 −5 −4 −3 −2 −1 0 1 2 3 4 5 6 x 4 3 2 1 −1 −2 −3 −4 A B

Describe fully the single transformation which maps triangle A onto triangle B. *(3 marks)*

target C

Rotations p. 100

Always write down the name of the transformation. This is a **rotation**. To fully describe a rotation you need to give (1) the amount of turn in degrees, (2) the direction (clockwise or anticlockwise), and (3) the coordinates of the centre of rotation.

TOP TIP

You need to describe a **single** transformation. You can't say that the shape is rotated then moved to the right. Use tracing paper to find a centre of rotation that maps triangle A directly onto triangle B.

Answers

UNIT 1 STATISTICS AND NUMBER

1. Place value

1 **(a)** 16 354 **(b)** forty thousand and thirty nine **(c)** 80

2 £974 £1497 £1749 £1947 £1974

2. Rounding numbers

1 **(a)** 3800 **(b)** 3760

2 **(a)** 64.9 **(b)** 64.89 **(c)** 64.893 **(d)** 60

3. Fractions

1 $\frac{9}{15}$ and $\frac{16}{24}$

2 £120

4. Calculator skills

1 **(a)** 91.125 **(b)** 5.256431913

2 6.474074074

5. Percentages

1 £546

2 35%

6. Percentage change 1

1 £14 310

2 34.6%

7. Ratio 1

1 270 m*l*

2 56

8. The Data Handling Cycle

Select, at random, 20 girls and 20 boys from your year group, and ask them to keep a record, over the next week, of how many times they log on to a social networking site.

Calculate an average (mean and/or median) for each group.

Write a conclusion making a statement about 'average usage' for both girls and boys and say whether the hypothesis is likely to be true or false.

9. Collecting data

(a) Discrete because the number of people owning a dog or a cat must be a whole number.

(b) Continuous because the amount of rainfall can take on any value, including decimals.
e.g. 720 m*l*, 841.6 m*l*, 29.2 inches.

10. Surveys

(a)

	Tally	Frequency
Buy lunch		
Bring from home		

(b) How much do you spend on lunch each day?
£0–£0.99 ☐ £1–£1.99 ☐ £2–£2.99 ☐ £3–£3.99 ☐
£4 or more ☐

11. Two-way tables

	Boys	Girls	Total
Left-handed	3	6	9
Right-handed	11	12	23
Total	14	18	32

12. Pictograms

(a) 9 **(b)** 30

(c)

Soaps	
Cartoons	
Films	
Sport	
Other	

Key ☐ represents 1 child

13. Bar charts

(a) English **(b)** 18

(c) PE **(d)** Maths

14. Measuring and drawing angles

(a) 28° +/− 2° **(b)** 307° +/− 2°

15. Pie charts

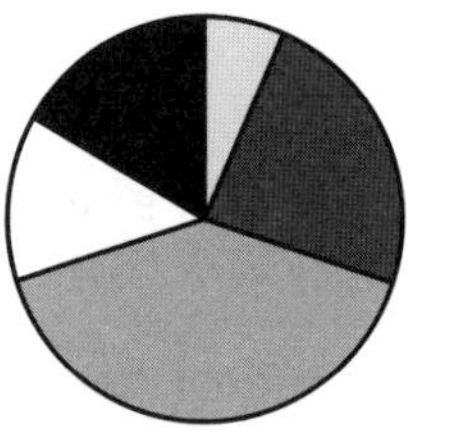

All sector sizes +/− 2°
Audi = 24° BMW = 84° Ford = 144°
Nissan = 48° Peugeot = 60°

16. Averages and range

(a) 14 **(b)** 7

(c) 13 **(d)** 12

17. Stem-and-leaf diagrams

(a)

Stem	Leaf
0	8 9 9
1	0 1 2 2 5 6 8 9
2	3 4 4 4 6 8

Key 1|2 represents 12 minutes

(b) 24 minutes **(c)** 16 minutes **(d)** 20 minutes

18. Averages from tables 1

(a) 1 goal **(b)** 2 goals **(c)** 95 ÷ 50 = 1.9 goals

19. Averages from tables 2

(a) $10 < x \leq 20$ **(b)** $20 < x \leq 30$ **(c)** 1350 ÷ 50 = 27

20. Scatter graphs

(a)

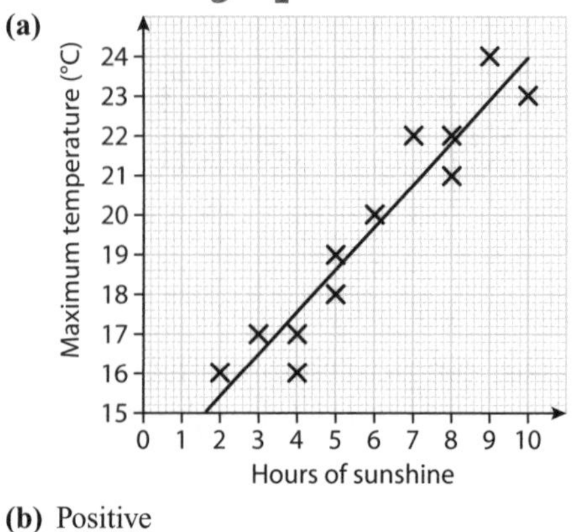

(b) Positive

(c) The greater the number of hours of sunshine, the greater the maximum temperature. There is a strong, positive correlation.
or,
As the number of hours of sunshine increased, so did the maximum temperature. There is a strong, positive correlation.

21. Frequency polygons

(a)

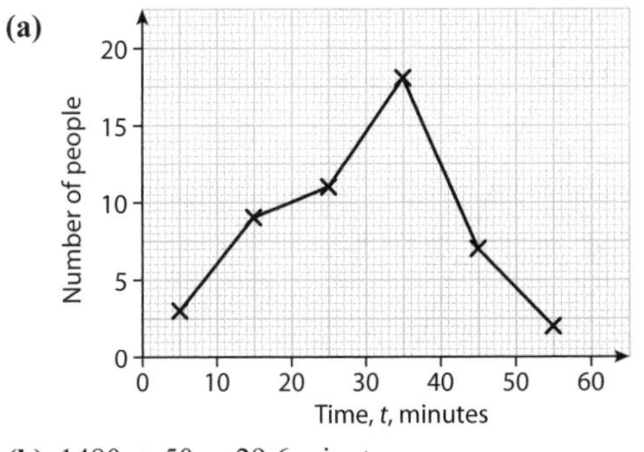

(b) $1480 \div 50 = 29.6$ minutes

22. Probability 1

(a) $\frac{1}{3}$ **(b)** $\frac{5}{6}$

23. Probability 2

(a) 0.7 **(b)** $0.3 \times 80 = 24$

24. Combinations

(a) $\frac{13}{24}$ **(b)** Red = 6 Blue = 16 Green = 26

25. Relative frequency

(a)

Number of trials	10	20	30	40	50	60
Number of yellow balls	3	7	12	17	23	27
Relative frequency of 'yellow'	0.3	0.35	0.4	0.425	0.46	0.45

(b)

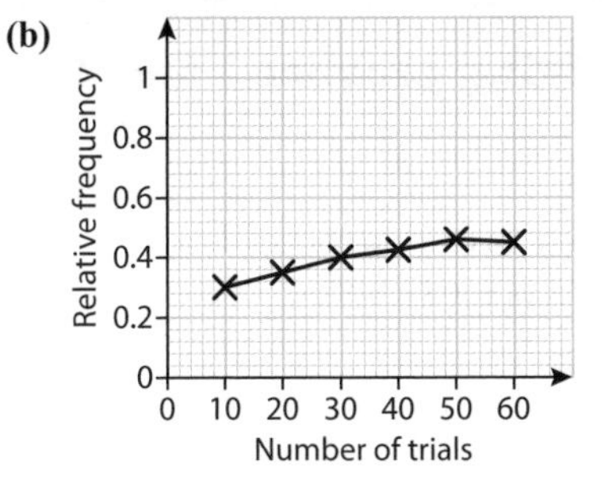

(c) Estimation: $0.45 \times 100 = 45$ (accept 42 to 48)

26. Comparing data

Anna's mean $= 104 \div 8 = 13$ Anna's range $= 16 - 9 = 7$
Carla's mean $= 11$ Carla's range $= 14 - 10 = 4$

On average, Anna scored better than Carla because she had a higher mean.

Carla's scores were more consistent than Anna's because she had a smaller range.

27/28. Problem-solving practice 1 and 2

1 10p

2

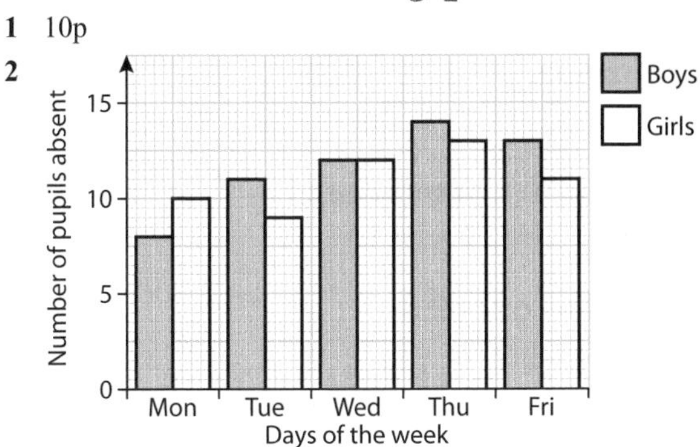

3 9

4 £223.56 for Business First, £208 for Leisure One, so "No, Callum is not".

5 0.33 seconds (2 d.p.)

UNIT 2 NUMBER AND ALGEBRA

29. Adding and subtracting

1 **(a)** 1714 **(b)** 462

2 £1.57

30. Multiplying and dividing

1 **(a)** 228 **(b)** 442 **(c)** 156 **(d)** 129

2 £70.72

3 7 tins, 59p change.

31. Decimals and place value

1 **(a)** 8 tenths **(b)** 8 thousandths

2 0.508 0.51 0.517 0.571 0.58

3 **(a)** 8.736 **(b)** 130

32. Operations on decimals

1 **(a)** 15.454 **(b)** 6.78 **(c)** 195.84 **(d)** 380

2 **(a)** £239.40 **(b)** 94p or £0.94

33. Decimals and estimation

1 **(a)** 3.813 **(b)** 155 **(c)** 10

2 4000

34. Negative numbers

1 **(a)** −15 **(b)** 48 **(c)** −5

2 $\boxed{-10} + \boxed{2}$ $\boxed{-9} + \boxed{1}$ $\boxed{-5} + \boxed{-3}$

35. Squares, cubes and roots

1 **(a)** 25 **(b)** 12

2 **(a)** 64 **(b)** 10

3 $3^4, 10^2, 5^3, 2^7$

36. Factors, multiples and primes

1 **(a)** 21 **(b)** 15 **(c)** 24 **(d)** 2 and 37

2 $2^3 \times 5 \times 7$

37. HCF and LCM

(a) $2 \times 2 \times 2 \times 3 \times 5$ **(b)** 60 **(c)** 360

38. Operations on fractions

(a) $\frac{19}{40}$ **(b)** $\frac{7}{55}$ **(c)** $\frac{1}{6}$ **(d)** $\frac{14}{15}$

39. Mixed numbers

1 **(a)** $5\frac{5}{6}$ **(b)** $2\frac{6}{25}$

2 $2\frac{1}{2} \div \frac{1}{3} = 7\frac{1}{2}$ so 7 full cups.

40. Fractions, decimals and percentages

1 **(a)** $\frac{3}{20}$ **(b)** $\frac{17}{25}$

2 $\frac{3}{10}$, (0.3), 36% (0.36), 0.42

3 $\frac{2}{5}$ (40%)

41. Percentage change 2

Cruks (£171.50 as against £175.50 at Spivs)

42. Ratio 2

£160

43. Collecting like terms

1 **(a)** $12t$ **(b)** $10h$ **(c)** $8w$

2 **(a)** $9d - 2k$ **(b)** $6y + 4m$ **(c)** $7x - 23$

44. Simplifying expressions

(a) n^5 **(b)** $6r^3$ **(c)** $32yz$
(d) $4f$ **(e)** 7

45. Indices

(a) m^9 **(b)** h^8 **(c)** y^{18}
(d) t^{20} **(e)** k^6 **(f)** a^5

46. Expanding brackets

1 **(a)** $3y - 18$ **(b)** $m^2 + 7m$
2 **(a)** $13a - 6b$ **(b)** $-w + 21$ **(c)** $m^2 + 22m$

47. Factorising

1 **(a)** $5(h + 3)$ **(b)** $k(k - 4)$ **(c)** $p(p + 1)$
2 **(a)** $4(2y - 3)$ **(b)** $5r(4r + 3)$ **(c)** $6n(1 - 3n)$

48. Equations 1

1 **(a)** $y = 8$ **(b)** $m = 17$ **(c)** $k = 19$
2 **(a)** $n = 5$ **(b)** $w = 10^1/_2$ **(c)** $a = 45$

49. Equations 2

1 **(a)** $t = -11$ **(b)** $h = 2^1/_2$
2 **(a)** $y = 14^1/_2$ **(b)** $m = -1$

50. Number machines

−3.5

51. Inequalities

1 **(a)** $x > -3$ **(b)** $x \leqslant 4$
2 **(a)**

−5 −4 3 2 1 0 1 2 3 x

(b)

−3 −2 −1 0 1 2 3 4 5 6 x

52. Solving inequalities

1 **(a)** $x \geqslant 5$ **(b)** $x < 4$
2 −4, −3, −2, −1, 0, 1, 2

53. Substitution

1 12
2 −14
3 6

54. Formulae

1 A = 36
2 $D = 49$

55. Writing formulae

1 $T = 50x + 35y$
2 $W = 800 + 150n$

56. Rearranging formulae

1 $p = \frac{3m}{n}$ 2 $t = \frac{k+7}{5}$ 3 $w = \frac{2-h}{6}$

57. Using algebra

1 $x + x - 5 + 2x = 31$, Tim's age $(x) = 9$
2 When $n = 5$, $n^2 + 4n - 3 = 42$ which is also equal to 6×7. $n = 12$ also works.

58. Coordinates

1 **(a)** D is (3, 6) **(b)** $AB = 6$ units **(c)** E is (6, 4)

59. Straight-line graphs 1

(a) $\frac{1}{2}$ **(b)** −3 **(c)** $\frac{3}{2}$ **(d)** $-\frac{1}{6}$

60. Straight-line graphs 2

(a)

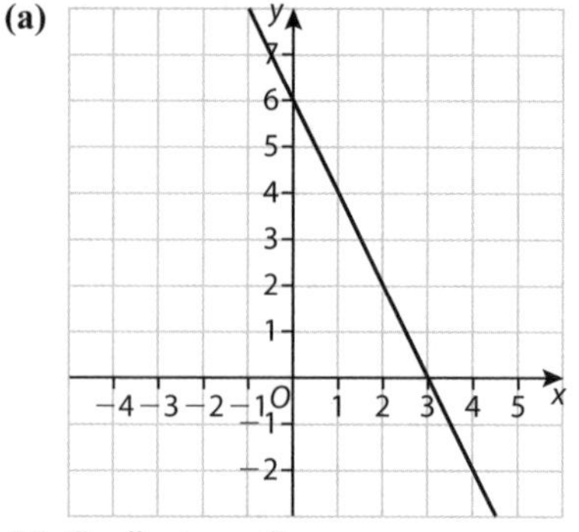

(b) Gradient = −2

61. Real-life graphs 1

(a) $C = 4d + 10$
(b) Miller's Tool Hire, Miller's = £26, Green's = £28

62. Distance-time graphs

(a) 4 minutes **(b)** 23 minutes **(c)** 0.4 miles

63. Sequences 1

1 **(a)** The difference between the terms increases by 2 each time.
(b) 33 and 45
2 **(a)** 31, 35, 39 **(b)** 19th term (the term is 103)

64. Sequences 2

(a) $3n - 1$ **(b)** 86 **(c)** 66
(d) No, 52nd term = 155, 53rd term = 158

65/66. Problem-solving practice 1 and 2

1 $\frac{30}{12}$ or $\frac{75}{30}$
2 **(a)** 14
(b) No. All the patterns use an even number of dots.
3 Sportscentre Trainers (£45 vs £60 vs £48)
4 $5x + 6 = 41$; $x = 7$
5

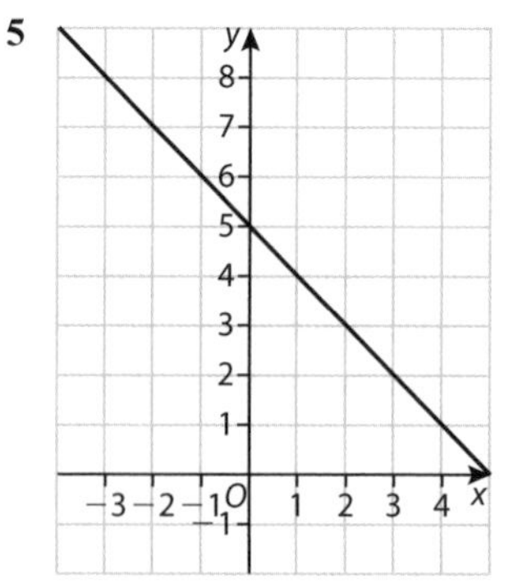

UNIT 3 GEOMETRY AND ALGEBRA

67. Proportion

1 £76.90 2 16 days

68. Trial and improvement

1 $x = 2.8$ 2 $x = 3.4$

69. Quadratic graphs

(a)

x	−2	−1	0	1	2	3	4
y	4	−1	−4	−5	−4	−1	4

(b), (c)

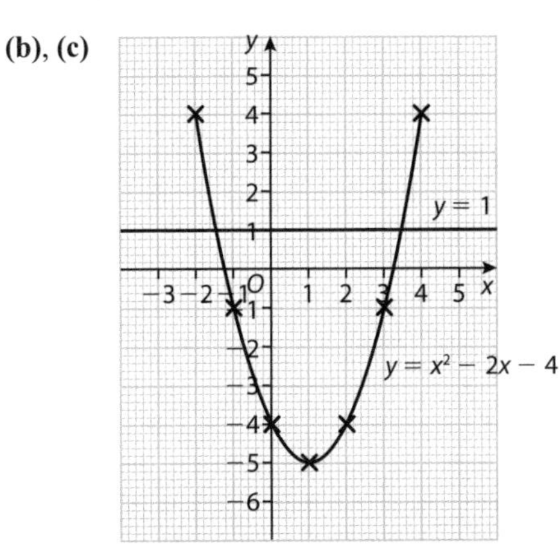

(d) From (−1.5, 1) to (−1.4, 1) **and** from (3.4, 1) to (3.5, 1)

70. Using quadratic graphs

(a) −0.2 and 4.2

(b) −1.5 and 5.5

71. Real-life graphs 2

(a) 13 pounds

(b) 8.2 kilograms

(c) 10 kg = 22 pounds
40 kg = 4 × 22 pounds = 88 pounds

72. Measuring lines

1 **(a)** 28 mm

(b) 55 mm

(c) 49 mm

(d) 60 mm

2 Surfboard 1.0 cm, Bus 4.7 cm, Real length of bus = 23.5 feet (accept 22 to 25 feet)

73. Metric units

1 **(a)** litres

(b) metres

(c) kilograms

(d) centimetres

2 1.7 mm

74. Angles 1

(a) 110°; angles around a point add up to 360°.

(b) 55°; vertically opposite angles are equal.

(c) 125°; angles on a straight line add up to 180°.

75. Angles 2

(a) 134°, corresponding angle **(b)** 97°

76. Solving angle problems

1 $x = 109°$

2 $y = 32°$

77. Angles in polygons

$x = 144°$

78. Symmetry

1 **(a) (i)** 1 **(ii)** 2 **(iii)** 0 **(iv)** 6

(b) (i) none **(ii)** 2 **(iii)** 2 **(iv)** 6

79. Quadrilaterals

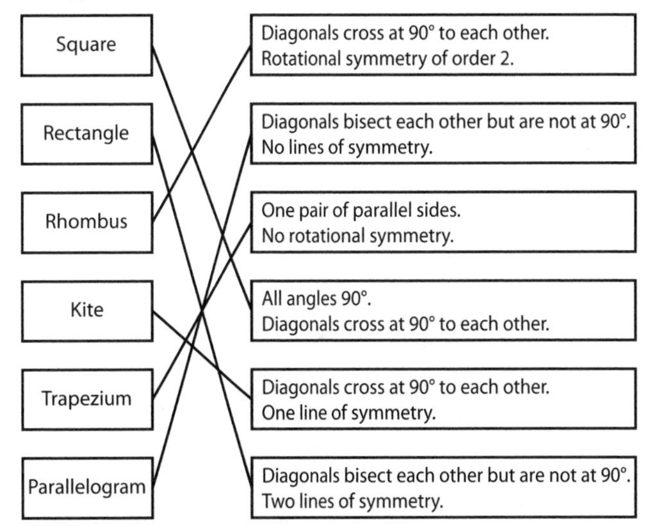

80. Perimeter and area

28 whole squares + 16 part squares = 36 units of area

81. Using area formulae

$\frac{1}{2} \times (8 + 13) \times 4 = 42\,\text{cm}^2$, $\frac{1}{2} \times 14 \times h = 42$, $h = 6\,\text{cm}$

82. Solving area problems

(a) $x = 3$ cm **(b)** $y = 5$ cm

(c) Area = 58 cm^2 **(d)** Perimeter = 50 cm

83. Circles

3.6 cm

84. Area of a circle

9.4 cm

85. 3-D shapes

30 cm^2

86. Volume

6.5 cm

87. Prisms

(a) 600 cm^3

(b) 2 × 30 + 100 + 240 + 260 = 660 cm^2

88. Volume of a cylinder

55 litres

89. Plans and elevations

(a) **(b)** **(c)**

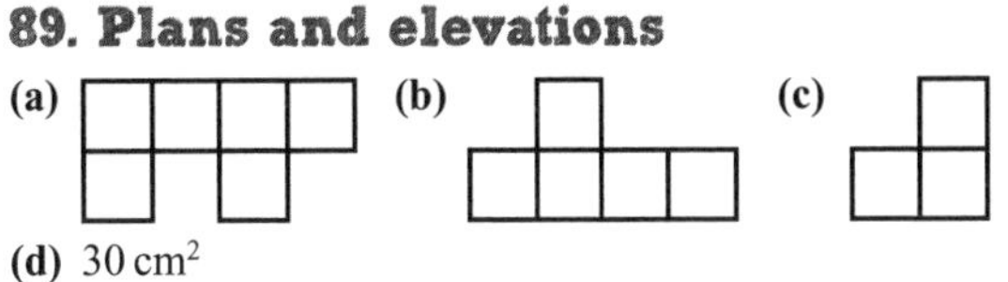

(d) 30 cm^2

90. Bearings

(a) 283 km (accept 275 km − 290 km)

(b) 225° (accept 223° − 227°)

(c)

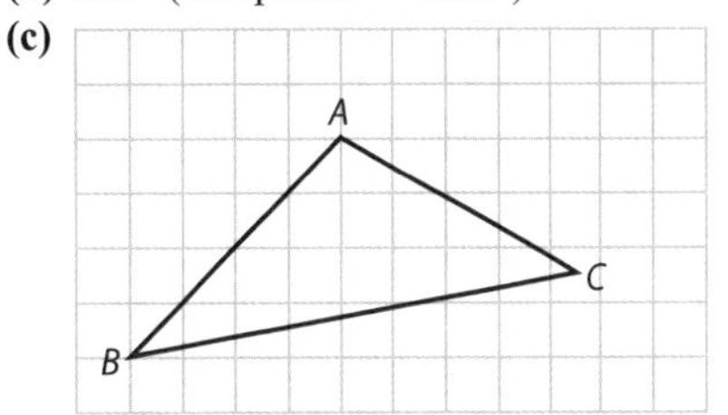

91. Scale drawings and maps

(a) 9 km

(b) 16 cm

92. Constructing bisectors

(a), (b)

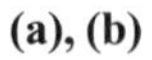

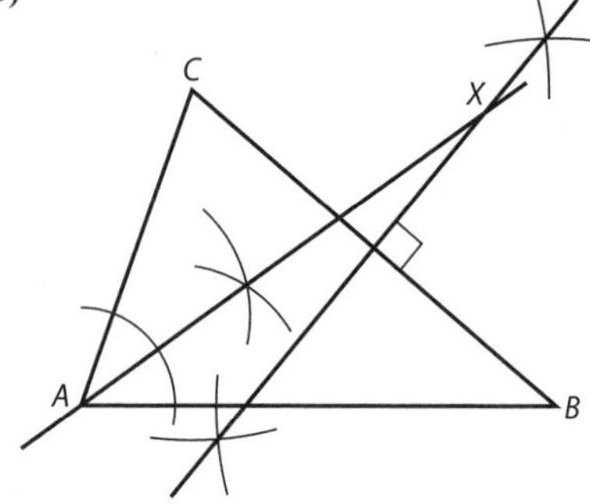

(c) 7.2 – 7.4 cm

93. Constructing triangles

(a) Accurate construction

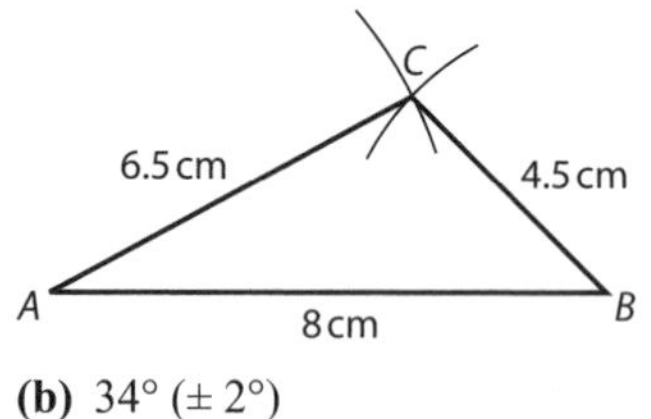

(b) 34° (± 2°)

94. Loci

(a) (b)

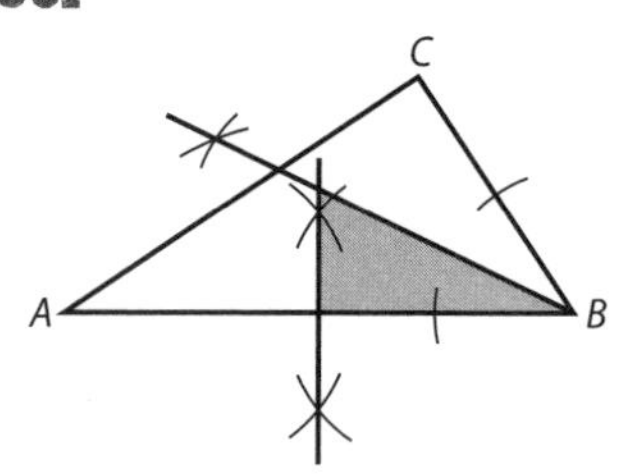

95. Speed

1 4 hours 15 minutes

2 266 miles

96. Measures

1 81.25 mph

2 12 gallons = 54 litres, £71.28

97. Congruent shapes

A and I B and G C and H E and J

98. Translations

(a)

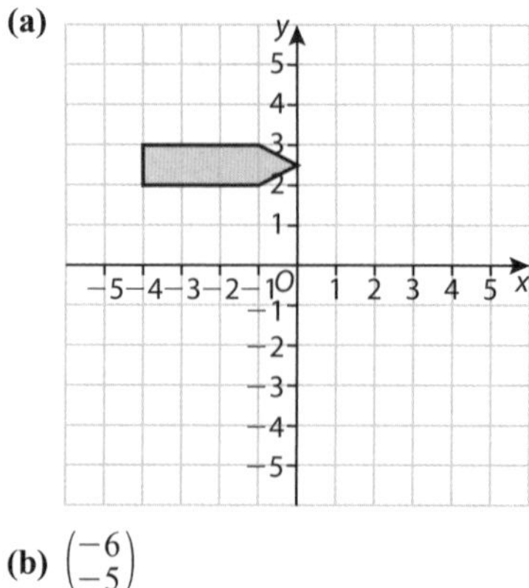

(b) $\begin{pmatrix} -6 \\ -5 \end{pmatrix}$

99. Reflections

(a), (b), (c)

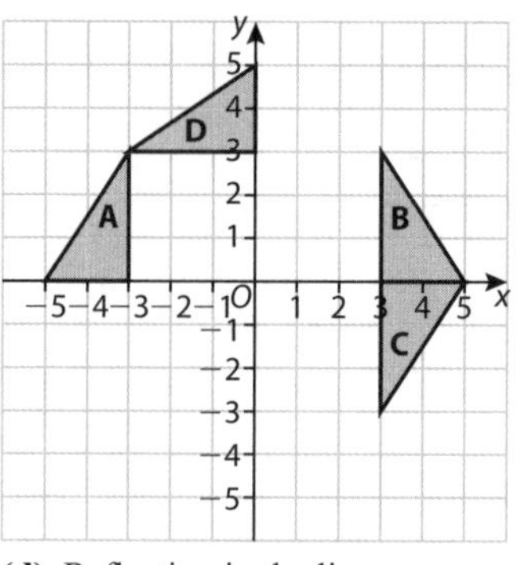

(d) Reflection in the line $y = -x$

100. Rotations

(a)

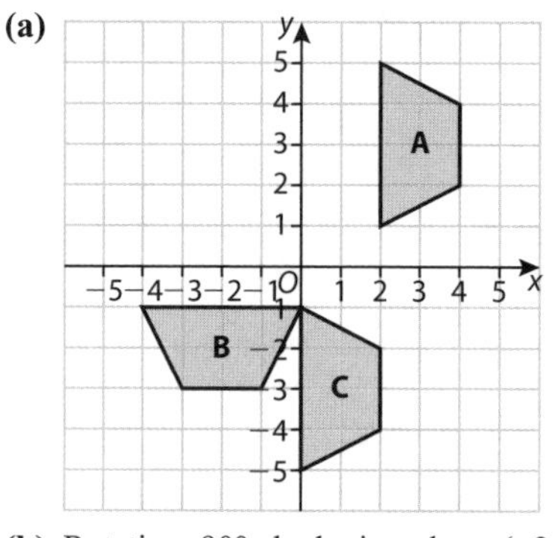

(b) Rotation, 90° clockwise, about (−2, 3)

101. Enlargements

(a) SF = 2

(b)

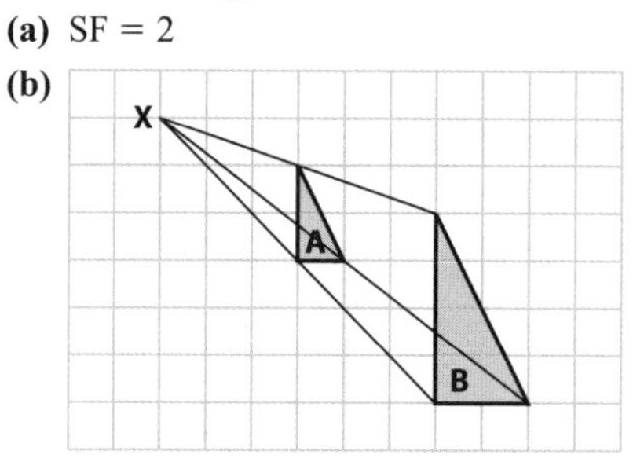

102. Similar shapes

A and K D and G H and J

103. Pythagoras' theorem

(a) $y = 3.5$ cm

(b) $z = 9.1$ cm

104/105. Problem-solving practice 1 and 2

1 e.g. Because 125° + 52° = 177° and the angles on a straight line should add up to 180°.

2 e.g.

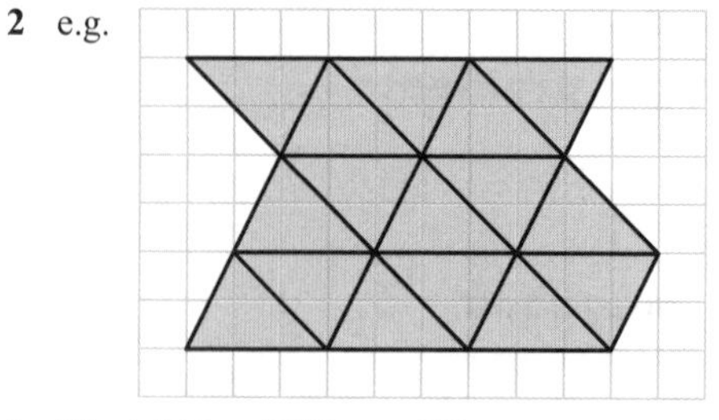

3 $10y + 90° = 360°$, $y = 27°$

4 13.8 km (1 d.p.)

5 Rotation 90° clockwise, centre (0, −4)

Published by Pearson Education Limited, Edinburgh Gate, Harlow, Essex, CM20 2JE.

www.pearsonschoolsandfecolleges.co.uk

Edited and produced by Wearset Ltd, Boldon, Tyne and Wear
Typeset and illustrated by Tech-Set Ltd, Gateshead
Cover illustration by Miriam Sturdee

First published 2013

16 15 14 13
10 9 8 7 6 5 4 3 2 1

British Library Cataloguing in Publication Data
A catalogue record for this book is available from the British Library

ISBN 978 1 447 94132 3

Printed in Slovakia by Neografia

In the writing of this book, no AQA examiners authored sections relevant to examination papers for which they have responsibility.

There are no questions printed on this page.

There are no questions printed on this page.

There are no questions printed on this page.

There are no questions printed on this page.

There are no questions printed on this page.

Revision is more than just this Guide!

You'll need plenty of practice on each topic you revise

1-to-1 page match with this Revision Guide.

Had a go ☐ Nearly there ☐ Nailed it! ☐ UNIT 3

Pythagoras' theorem

C Guided

1 Work out the length of AC.

Using Pythagoras' theorem $a^2 + b^2 = c^2$

$35^2 + 25^2 = AC^2$

$1225 + 625 = AC^2$

.... $= AC^2$

$AC = \sqrt{\ldots\ldots}$

$AC =$ ....

25 cm, 35 cm, A, B, C

Write down your whole calculator display before rounding your answer to a suitable degree of acuracy.

Answer .... cm (3 marks)

C Guided

2 The screen size of a television is the length, to the nearest inch, of the diagonal of the screen.

20 inches, 42 inches

Sketch and label a right-angled triangle and write on it the two dimensions given.

The screen on this 42 inch television is 20 inches high.
How wide is the screen? Give your answer to the nearest inch.

Using Pythagoras' theorem $a^2 + b^2 = c^2$

$BC^2 + 20^2 = 42^2$

A, B, C, 42 inches, 20 inches

Answer .... inches (3 marks)

C

3 Two mobile phone masts are 2.1 km apart on horizontal ground.
Work out the distance between the tops of the masts.
Give your answer to the nearest metre.

25 m, 85 m, 2.1 km

Sketch and label a right-angled triangle.

Answer .... m (4 marks)

103

Exam-style questions on this topic.

Guided questions help build your confidence.

Grades, marks and hints get you well prepared for this topic in your exam.

Check out the matching Workbook!

THE REVISE SERIES FROM PEARSON

www.pearsonschools.co.uk/reviseaqa